AF566891

Elliptic Cohomology

THE UNIVERSITY SERIES IN MATHEMATICS

Recent volumes in the series:

THE CLASSIFICATION OF FINITE SIMPLE GROUPS: Volume 1: Groups of Noncharacteristic 2 Type
Daniel Gorenstein

COMPLEX ANALYSIS AND GEOMETRY
Edited by Vincenzo Ancona and Alessandro Silva

ELLIPTIC COHOMOLOGY
Charles B. Thomas

ELLIPTIC DIFFERENTIAL EQUATIONS AND OBSTACLE PROBLEMS
Giovanni Maria Troianiello

FINITE SIMPLE GROUPS: An Introduction to Their Classification
Daniel Gorenstein

AN INTRODUCTION TO ALGEBRAIC NUMBER THEORY
Takashi Ono

MATRIX THEORY: A Second Course
James M. Ortega

PROBABILITY MEASURES ON SEMIGROUPS: Convolution Products, Random Walks, and Random Matrices
Göran Högnäs and Arunava Mukherjea

RECURRENCE IN TOPOLOGICAL DYNAMICS: Furstenberg Families and Ellis Actions
Ethan Akin

TOPICS IN NUMBER THEORY
J. S. Chahal

VARIATIONS ON A THEME OF EULER: Quadratic Forms, Elliptic Curves, and Hopf Maps
Takashi Ono

Elliptic Cohomology

Charles B. Thomas
University of Cambridge
Cambridge, England

Springer Science+Business Media, LLC

Library of Congress Cataloging-in-Publication Data

Thomas, C. B. (Charles Benedict)
Elliptic cohomology / Charles B. Thomas.
p. cm. -- (The university series in mathematics)
Includes bibliographical references and index.

DOI 10.1007/978-0-306-46969-5

1. Homology theory. I. Title. II. Series: University series in mathematics (Plenum Press)
QA612.3.T48 1999
514'.23--dc21
98-45731
CIP

Originally published by Kluwer Academic/Plenum Publishers in 1999
MyCopy version of the original edition 1999

10 9 8 7 6 5 4 3 2 1

A C.I.P. record for this book is available from the Library of Congress.

Preface

Elliptic Cohomology was developed in the mid-1980s following the discovery that a multiplicative genus defined on oriented cobordism, localized away from the prime 2, has a surprising relation to Jacobi elliptic functions. After an initial flurry of activity, interest waned in the new theory, partly because of slowness in the appearance of applications but mostly because of the lack of a satisfactory bundle-theoretic model. In one direction at least, the situation is now more promising — moonshine phenomena of interest to finite group theorists provide the means to give a usable geometric definition at least for the classifying space of a finite group, and repaying the compliment, elliptic cohomology appears to offer a good framework in which to formulate moonshine for a preferred family of simple groups. As a stimulus to further activity, it seems right to me to attempt to give a survey, however provisional, of what is currently known. It has been plausibly claimed also that one variant of the theory captures much of what is known about the stable homotopy groups of spheres; it is my own hope that with more geometric input, elliptic cohomology may resolve some of the open questions, which seem just beyond the reach of K-theory.

This survey is an expanded version of lectures that I originally gave at Ohio State University in the spring quarter, 1993. These have circulated for some time in handwritten form, and I hope that the present version is both clearer and more complete. Prerequisites are a good knowledge of algebraic topology and the theory of finite groups. Some acquaintance with the theory of modular forms will also be of assistance. In writing the book I have increasingly realized that it is a survey in the sense that I have attempted to give the flavor of what is known and that in many places I have referred the reader to more detailed texts. In a subject that straddles several areas of mathematics, this may be inevitable, but if the book encourages further research along the lines I have indicated, I will be more than content.

As with my earlier book on the cohomology of finite groups, this book is dedicated with love and gratitude to my wife, Maria.

Cambridge, September 1998.

Contents

Elliptic Cohomology

Introduction

Elliptic cohomology combines ideas and results from algebraic topology, arithmetic, representations of finite groups and theoretical physics in the shape of topological field theories. From the algebraic point of view, it is a quotient of spin cobordism, and as such it forms part of a chain:

$$\Omega^*_{\text{spin}} \to \cdots ? \to Ell^* \to K\text{Spin}^* \to H^*,$$

with each link corresponding to a 1-dimensional commutative formal group law. In the case of elliptic cohomology this was written down by Euler in the eighteenth century, and the validity of Eilenberg–Steenrod axioms for the corresponding cohomology theory follows from properties of addition on a class of elliptic curves in characteristic p. Universality of the formal group law for cobordism implies that there is a corresponding ring homomorphism or genus taking values in the coefficients of the theory concerned.

In 1988 G. Segal gave a talk in the Bourbaki Seminar [101] in which he summarized what was known at the time under the headings:

- $Ell^*(X)$ is a cohomology theory [69, 70].
- The structural genus is rigid with respect to compact, connected group actions [25].
- The completed localized ring $Ell^*(BG)^\wedge_p$ is determined by elliptic characters [56].

Segal also explained that a genus related to the universal elliptic genus should be defined as the index of a Dirac operator in infinite dimensions and suggested that a geometric model for $Ell^*(X)$ could be constructed by using ideas from conformal field theory.

We devote Chaps. 1–3 to a survey of this material. The first two headings are related, since we can use rigidity to prove the exactness of $Ell^*(X)$. As Segal elegantly explains rigidity of the universal elliptic genus φ is equivalent to its strong multiplicativity, that is, to the result $\varphi(E) = \varphi(F)\varphi(B)$ for any fibration $p : E \to B$ by spin manifolds having a compact connected structural group. The genus φ is a homomorphism from oriented cobordism (localized away from 2) into the polynomial ring $\mathbb{Z}[1/2][\delta,\varepsilon]$, where δ and ε can be interpreted as modular forms of weights 4 and 8, respectively, and these are targeted by manifold classes represented by $\mathbb{C}P^2$ and $\mathbb{H}P^2$. Using the strong multiplicative property, we can choose higher dimensional generators $x_{12}, x_{16}, \ldots$ for $\Omega^*_{SO} \otimes \mathbb{Z}[1/2]$ generating an ideal containing all classes on which the genus φ takes the value 0. Dividing by this ideal corresponds geometrically to allowing bordism of manifolds with singularities of a prescribed type (see [17] for a careful explanation of this), so that we obtain a connective homology theory with coefficients in $\mathbb{Z}[1/2][\delta,\varepsilon]$. Inverting either ε or the discriminant Δ leads to a periodic cohomology theory.

The elliptic character ring for a finite group G was introduced in [56] as part of the authors' generalization of M. Atiyah's completion theorem for $K(X)$ (a v_1-periodic theory) to arbitrary v_n-periodic theories. Roughly speaking an elliptic class function is defined on conjugacy classes of pairs of commuting elements in G, and it takes values in a sufficiently large extension of the coefficient ring of modular forms. In its original form (see Chap. 3, the main theorem of [56] required localization at a single prime p followed by completion. But their work made clear that if we make the right technical assumptions (such as allowing certain types of denominator), a representation-theoretic description of $Ell^*(BG)$ was possible. Recent progress in this direction is the major theme of part 2 of this work.

Calculating $H^*(BG,\mathbb{Z})$, and in particular determining the Chern subring, follows from properties of the Mackey functor:

$$K \mapsto H^*(BK,\mathbb{Z}) \qquad (K \leqslant G).$$

These include Frobenius reciprocity and the double-coset formula, which are shared by more general complex-oriented cohomology theories. The best-behaved is $K^*(BG)$, since graded by $\mathbb{Z}/2$ we have $K^{\text{even}}(BG) \cong R(G)^\wedge$ and $K^{\text{odd}}(BG) = 0$. Problems created by $H^{\text{odd}}(BG,\mathbb{Z})$ have disappeared. If we replace K^* by Ell^*, something curious happens: The increased richness of the coefficients still removes classes coming from H^{odd}, but it also introduces torsion of a new kind into Ell^{odd}. Happily the latter problem does not arise for finite groups with a restricted Sylow subgroup structure, and outside the primes 2 and 3 for example, all but four of the sporadic simple groups fall into this class. Defining the elliptic character ring $\mathcal{E}ll^*_G$ as proposed by Devoto, we can then prove an analog of Atiyah's theorem if we tensor with $\mathbb{Z}[1/|G|]$. We can avoid this requirement by restricting attention to groups G of odd order such that each Sylow subgroup G_p has order p^t with $t \leqslant 4$.

Corresponding to an elliptic character is an elliptic object. This is an infinite-dimensional bundle over the loop space LBG with a grading into finite-dimensional flat subbundles induced by the Diff (S^1)-action on the fibers, restricted to the rotational subgroup. Such objects fit naturally into the framework for conformal field theory developed by Segal already mentioned. From a quite different direction, the 2-variable Thompson series forming one part of the moonshine data for certain finite simple groups admits interpretation as an elliptic object. In the case of the large Mathieu group M_{24}, a first candidate for a moonshine module is obtained by looking for an element in $Ell^*(BM_{24})$ invariant under the action of the Hecke subalgebra of cohomology operations. This element exists because numerical conditions under which a one-dimensional Hecke eigenspace of modular forms exists can a posteriori be determined from the cycle types of conjugacy classes of elements in $M_{24} \subset S_{24}$. This numerology provides some evidence for J. MacKay's suggestion that moonshine is interesting only for the groups M_{24}, Co_0, and $\mathbb{M}$. (The systematic construction of an umbrella for elliptic objects coming from various sources forms part of an ongoing collaboration between the author and A. Baker.)

In modifying the first Thompson series for M_{24} to satisfy the genus zero conditions for a true moonshine, the prime 2 behaves quite differently to odd primes dividing the order. This is one indication that a new construction is needed to extend the definition of elliptic cohomology to the prime 2. This has been achieved *homologically* by Kreck and Stolz, who set

$$Ell_*(X) = \bigoplus_{k \geqslant 0} \Omega^{\mathrm{Spin}}_{*+8k}(X)/\sim,$$

where $\sim$ identifies the total space of an $(\mathbb{H}P^2, PSp_3)$-fibration with its base. Localized at the prime 2, the spectrum for elliptic cohomology is homotopy equivalent to $\bigvee_{k \geqslant 0} \Sigma^{8k} ko$ with a product structure other than that coming from multiplication in ko. This result is interesting for several reasons:

1. Elliptic cohomology can be naturally defined over $\mathbb{Z}$ rather than $\mathbb{Z}[1/2]$. This is necessary for a bundle-theoretic definition along the lines of K-theory as was partially achieved in the special case of $X = BG$.

2. The anomalous behavior of the prime 2 in constructing a moonshine module satisfying the full list of Conway–Norton conditions may have a plausible explanation at least for the group M_{24}.

3. The Kreck–Stolz homology theory is one of a sequence $Ell^{\mathbb{R}}_*(X)$, $Ell^{\mathbb{C}}_*(X)$, $Ell^{\mathbb{H}}_*(X)$ to which we would like to add $Ell^{\mathbb{O}}_*(X)$, where spin is replaced by the group $O\langle 8\rangle$ and $\mathbb{H}P^2$ by the Cayley projective plane $\mathbb{O}P^2$. As yet such a (co)homology theory has to be shown to exist, although reasons other than its putative definition suggest that it does. These are discussed in Chap. 10.

Heuristically we reason that $Ell_*^{\mathbb{O}}(X)$ should be related to complex abelian surfaces in the same way that $Ell_*^{\mathbb{H}}(X)$ is related to elliptic curves. Such theories are constructed with formal group laws having heights h, where $1 \leqslant h \leqslant 10$. The theory so far suggests that $h = 4$ is the most interesting, although as yet we looked only at specific examples where height equals 2, that is, at theories providing no better information than $Ell_*^{\mathbb{H}}$. The hope is to construct a v_4-periodic theory as a quotient of $\Omega_{\langle 8\rangle}^*(X)$ with both 2 and 3 invertible in a coefficient ring having generators identifiable with certain Siegel modular forms of degree 2.

Additional evidence that such a theory (or theories) should exist is provided by the spectrum eo_2 being studied by Hopkins, Mahowald, Strickland, and others. In this setting the Witten genus, regarded as a structural map from $\Omega_*^{\langle 8\rangle} \to \mathbb{Z}[1/6][g_2, g_3]$, should be the specialization of a map into power series in 2 variables (restrictions of Siegel modular forms?) in much the same way that the $\hat{A}$-genus is a specialization of a map $\Omega_*^{\mathrm{Spin}} \to \mathbb{Z}[1/2][\delta, \varepsilon]$ (take $\delta = -1/8$, $\varepsilon = 0$).

Since its introduction in the mid-1980s elliptic cohomology has been a beautiful theory looking for applications. To paraphrase Turgenev on Russian* — "surely such a language has been given for a great purpose." These are slowly beginning to appear, we give evidence that elliptic objects provide the correct setting for moonshine phenomena; related examples occur in [46]. Hopkins claimed that the spectrum eo_2 encodes the stable homotopy groups of spheres in dimensions up through 60. More generally we hope that it may help to settle questions about the Adams spectral sequence just outside the reach of K-theory. Geometrically there is evidence of a relation between the vanishing of genera of elliptic type and the existence of metrics with positive curvature properties. And once we have more calculations at our disposal — particularly at the prime 2 — it should be possible to use characteristic classes taking values in $Ell^*(X)$ to refine other obstructions at present expressed in terms of $H^*(X, \mathbb{Z})$ or $K^*(X)$.

This is a very partial account of elliptic cohomology from earlier work on the ordinary cohomology of discrete groups and a long-standing interest in the 26 sporadic simple groups. Through ignorance we included little on stable homotopy theory and nothing on the motivation from theoretical physics. The book is based on notes for a course of lectures given at Ohio State University in 1993 and later repeated in shortened form at Stanford, Essen, and in Barcelona. Chapter 3 is based on notes of John Hunton on the various preprint versions of [56], Chap. 8 represents joint work with Andrew Baker, and for Chap. 10 we are deeply indebted to various exchanges with Jack Morava. We are also grateful to Jorge Devoto for making his work available at an early stage and to Henry Glover for his steady interest and careful questioning in our lectures.

We wish to acknowledge financial support under the Human Capital and Mobility program (HCM) funded by the European Union, from the previously

*Нелзя верит, чтобы такой язык не был дан великому народу. [118]

mentioned universities, and from the ETH in Zürich. Thanks are also due to Michèle Bailey here in Cambridge for her excellent typing of various versions of the manuscript; to Sylvain Cappell, who suggested its publication, and to the editorial staff at Plenum for their interest, care, and patience.

1

Elliptic Genera

1.1. Oriented Cobordism Ring Ω_{SO}^*

Let W be a compact, oriented, differentiable $(n+1)$ manifold with boundary $V = \partial W$. Give V the orientation from W.

DEFINITION. Two manifolds V_1 and V_2 are said to be *cobordant* $(V_1 \sim V_2)$ if there exists a manifold W with $\partial W = V_1 \sqcup (-V_2)$. Here the symbol $\sqcup$ denotes disjoint union and, $-V$ is V with reversed orientation. It is not hard to see cobordism is an equivalence relation, the set of equivalence classes Ω_{SO}^n forms an abelian group under the operation $\sqcup$, and the Cartesian product makes $\Omega_{SO}^* = \sum_{n=0}^{\infty} \Omega_{SO}^n$ into a graded ring. This is anticommutative in the sense that $[U^m][V^n] = (-1)^{mn}[V^n][U^m]$, and it has a unit 1 = [point].

THEOREM 1.1. *$\Omega_{SO}^n \otimes \mathbb{Q} = 0$ unless $n \equiv 0 \pmod 4$ and Ω_{SO}^{4k} is a finitely generated abelian group with $\mathbb{Q}$-rank equal to the number of partitions of k.*

This result is due to R. Thom, who also introduced the notion of a *basic sequence* of manifolds $\{M^1, M^2, \ldots M^k \ldots : \dim M^k = 4k\}$ to describe polynomial generators of Ω_{SO}^*. One example of such a sequence is $\{\mathbb{C}P^2, \mathbb{C}P^4. \cdots, \mathbb{C}P^{2k}, \ldots\}$, the cobordism classes of products are distinguished by their Pontrjagin numbers (see Remark) and have dimensions that generate all partitions of k.

REMARK. $\mathbb{C}P^{2k+1}$ bounds, i.e., is cobordant to zero. Consider the locally trivial fiber bundle:

$$\begin{array}{ccc} \mathbb{C}P^1 = S^2 & \longrightarrow & \mathbb{C}P^{2k+1} \\ & & \downarrow \\ & & \mathbb{H}P^k \end{array}$$

The total space bounds the total space of an associated $SO(3)$ bundle with fiber D^3. The odd-dimensional projective spaces are important in characterizing the universal elliptic genus that follows.

We assume that the reader has some knowledge of Chern classes of complex vector bundles E over X or complex representation modules. Thus for each bundle E, there exist classes $c_i(E) \in H^{2i}(X,\mathbb{Z})$ satisfying

1. $c_o(E) = 1$, $c_i(E) = 0, i >$ fiber dimension of E.

2. $c_i(f^! E) = f^*(c_i E)$ for the pullback of E along $f : Y \to X$.

3. $c.(E_1 \oplus E_2) = c.E_1 c.E_2$, where $c. = c_0 + \cdots + c_n$ denotes the total Chern class.

4. $c.(H) = 1 - t$. Here H is the Hopf bundle over $\mathbb{C}P^n$, having as fiber over each projective point the line in $\mathbb{C}^{n+1}$ defining that point, and $t \in H^2(\mathbb{C}P^n,\mathbb{Z})$ is dual to the homology class carried by $\mathbb{C}P^{n-1}$.

One way of constructing the classes c_i is to proceed as follows. The ring $H^*(\mathbb{C}P^n,\mathbb{Z})$ is a truncated polynomial ring $\mathbb{Z}[t]/\langle t^{n+1}\rangle$, and a line bundle E over a CW-complex X is classified by a map $f : X \to \mathbb{C}P^\infty = \bigcup_n \mathbb{C}P^n$. This means that E is isomorphic as a vector bundle to the induced or pullback bundle $f^! H$. Set $c.(E) = 1 - f^* t \in H^{\text{even}}(X,\mathbb{Z})$, and we can extend the definition to a sum of line bundles over X, $E_1 \oplus \cdots \oplus E_m$, classified by a map into $\mathbb{C}P^\infty \times \cdots \times \mathbb{C}P^\infty$ (m times). Thus $c.(E_1 \oplus \cdots \oplus E_m) = \prod_{j=1}^m (1 - f_j^* t)$. The extension to an arbitrary bundle E depends on the so-called splitting principle; namely, given an arbitrary vector bundle E over X, there exists a continuous map $f : Y \to X$, such that:

- $f^!(E) \cong E_1 \oplus \cdots \oplus E_m$, a sum of line bundles.

- $f^* : H^*(X,\mathbb{Z}) \to H^*(Y,\mathbb{Z})$ is injective.

If BU_m denotes the classifying space for general m-dimensional bundles, then $H^*(BU_m,\mathbb{Z})$ can be identified with the ring of symmetric polynomials in $H^*(\mathbb{C}P^\infty \times \cdots \times \mathbb{C}P^\infty,\mathbb{Z}) \cong \mathbb{Z}[t_1,\ldots,t_m]$, and the elementary symmetric polynomials $y_1 \cdots y_m$ are called the universal Chern classes.

The manifold V^{2n} is said to be almost complex if its tangent bundle TV is obtained from a $\mathbb{C}^n$-bundle over V by forgetting the complex structure. By considering the S^1-equivariant embedding of S^{2n+1} in $\mathbb{R}^{2n+2}$, it is easy to see that $(T\mathbb{C}P^n) \oplus (1)$ is determined by $(n+1)$ copies of H and $c.(T\mathbb{C}P^n) = (1+t)^{n+1}$ (see [86, Sec. 14]).

In general we write $c.(V) = c.(TV)$ for an arbitrary almost complex manifold V.

DEFINITION. If E is a real vector bundle over X define the Pontrjagin classes $p_i(E)$ by:

$$p_i(E) := (-1)^i c_{2i}(E \otimes \mathbb{C}) \in H^{4i}(X, \mathbb{Z}).$$

The preceding properties for the Chern classes and the observation that $2c_{2i+1}(E) = 0$, imply

1. $p_0(E) = 1, p_i(E) = 0, 2i >$ fiber dimension of E.
2. $p_i(f^! E) = f^*(p_i E)$.
3. $p.(E_1 \oplus E_2) = p.E_1 \cdot p.E_2$ (modulo 2-torsion).
4. $p.(H_{\mathbb{R}}) = 1 + t^2$. Here $H_{\mathbb{R}}$ denotes the underlying real bundle for H, so that $H_{\mathbb{R}} \otimes \mathbb{C} = H \oplus \overline{H}$.

Generalizing Property 4 to the real tangent bundle of $\mathbb{C}P^n$, we have $p.(\mathbb{C}P^n) = p.(T\mathbb{C}P^n)$ is determined by $(1+t)^{n+1}(1-t)^{n+1}$, so that modulo our sign convention for the p_i, $p.(\mathbb{C}P^n) = (1+t^2)^{n+1}$.

DEFINITION. Let X be a compact, oriented, differentiable manifold of dimension $4n$ and $(i_1, \ldots, i_r)$ a partition of n. The *Pontrjagin number* corresponding to this partition equals

$$\left(\prod_{j=1}^{r} p_{i_j}(X)\right)[X],$$

where $[X] \in H_{4n}(X, \mathbb{Z})$ denotes the orientation class. Chern numbers are defined analogously.

EXERCISE. Use the preceding information to prove the claim that complex projective spaces $\mathbb{C}P^{2k}$ give a basic sequence for $\Omega^*_{SO} \otimes \mathbb{Q}$ (see [86, Sec. 16] for details).

Later on we refer to more special cobordism rings, for example Ω^*_{Spin} (for which we assume that tangent bundles admit a spin structure) and $\Omega^*_{\langle 8 \rangle}$ (for which we further assume that at least rationally, p_1 vanishes). In the case of spin cobordism $\Omega^*_{\text{Spin}} \otimes \mathbb{Q} \cong \Omega^*_{SO} \otimes \mathbb{Q}$, which obscures the delicate structure at the prime 2 for both theories.

1.2. Genera

DEFINITION. Let R be an integral domain over $\mathbb{Q}$. Then a *genus* is a ring homomorphism $\varphi : \Omega^*_{SO} \otimes \mathbb{Q} \to \mathbb{R}$ with $\varphi(1) = 1$.

There is a (1-to-1) correspondence between genera φ and even power series $Q(x) = 1 + a_2x^2 + a_4x^4 + \cdots$ in a variable x of weight 2. To see this consider the symmetric expression:

$$Q(x_1)\cdots Q(x_n) = 1 + a_2 \sum_{i=1}^{n} x_i^2 + \cdots,$$

where the term of weight $4r$ can be expressed as a homogeneous polynomial $K_r(p_1,\ldots,p_r)$ in the elementary symmetric functions p_j in the variables x_i^2. Each K_r has weight $4r$, it is independent of n for $1 \leqslant r \leqslant n$, and it is part of the *multiplicative sequence* determined by $Q(x)$.

THEOREM 1.2. *If the functions p_j are identified with the Pontrjagin classes of the tangent bundle TM, then the function:*

$$\varphi_Q(M) := K_r(p_1,\ldots,p_r), \qquad r \equiv 0 \ (\mathrm{mod}\ 4)$$
$$\varphi_Q(M) := 0,$$

otherwise it defines a genus.

Proof: The additivity if φ_Q is clear. Since the relation $p_. = p'_. p''_.$ implies that $\sum_{n\geqslant 0} K_n = \sum_{n\geqslant 0} K'_n \sum_{n\geqslant 0} K''_n$ by the multiplicative property of the K_n, φ_Q preserves Cartesian products. Finally let $M^{4n} = \partial W$, so that $TW|M = TM \oplus (1)$. Here (1) denotes the trivial normal bundle. The Pontrjagin classes of M are those of W restricted to the boundary, and an application of Stoke's theorem suffices to show that the Pontrjagin numbers of M are therefore all trivial. ■

To see that every genus φ arises from a power series in this way, we introduce the two following definitions.

DEFINITION. The logarithm $g(y) = \log_\varphi(y)$ associated with φ is given by the formal power series:

$$g(y) = \sum_{n=0}^{\infty} \frac{1}{2n+1} \varphi(\mathbb{C}P^{2n}) y^{2n+1}.$$

Note: $g(y) = y \ (\mathrm{mod}\ y^2)$ and $g(-y) = -g(y)$.

DEFINITION. The formal group law $F(x,y)$ with logarithm g is given by:

$$F(x,y) = g^{-1}(g(x) + g(y)).$$

It follows from the definition of F that:

$$F(x,0)=x, \qquad F(0,y)=y,$$
$$F(F(x,y),z)=F(x,F(y,z))$$
$$F(x,y)=F(y,x).$$

And F is *odd* in the sense that $F(-x,-y)=-F(x,y)$.

The logarithm g determines F; the converse is an exercise in formal partial differentiation. Thus if we write $g(F(x,y))=g(x)+g(y)$:

$$g'\cdot\frac{\partial F}{\partial y}=g'(y).$$

Evaluating at $(x,0)$ and noting that $F(x,0)=x$, $g'(0)=1$, we have

$$\frac{\partial F}{\partial y}(x,0)=\frac{1}{g'(x)}.$$

Let f be the formal inverse to the logarithm g, with g associated to an arbitrary genus φ.

LEMMA 1.1. *If the power series $Q(x)$ is defined by $Q(x):=x/f(x)$, then $g'(y)=d(\log_{\varphi_Q}(y))/dy$, i.e., the arbitrary genus φ and the genus φ_Q have the same logarithm.*

Sketch of proof: From the discussion of characteristic classes, it follows that $p.(\mathbb{C}P^n)=(1+p_1)^{n+1}$, hence $K(p_1,\dots,p_n)=K(p_1)^{n+1}=Q(x)^{n+1}$, with $p_1=x^2$, therefore:

$$\begin{aligned}
\varphi_Q(\mathbb{C}P^n) &= \left(\frac{x}{f(x)}\right)^{n+1}[\mathbb{C}P^n]\\
&= \text{coefficient of } x^n \text{ in } \left(\frac{x}{f(x)}\right)^{n+1}\\
&= \begin{array}{l}\text{applying the calculus of residues}\\ \text{for a suitable contour } \gamma\end{array} \quad \frac{1}{2\pi i}\int_\gamma\left(\frac{z}{f(z)}\right)^{n+1}dz\\
&= \frac{1}{2\pi i}\int_{f(\gamma)}\frac{1}{w^{n+1}}g'(w)\,dw\\
&= \text{coefficient of } y^n \text{ in } g'(y).
\end{aligned}$$

By the definition $g(y)=\log_\varphi(y)$ this coefficient equals $\varphi(\mathbb{C}P^n)$. ■

The important thing to note in the preceding discussion is that given any one of the four quantities $\varphi, Q, g=\log_\varphi$ and $F=F_\varphi$, we can determine the other three.

Now for some examples:

EXAMPLE 1.1.

$$\text{Let } Q(x) = \frac{x}{\tanh(x)}.$$

Then $f'(x) = \operatorname{sech}^2(x) = 1 - \tanh^2(x)$. Therefore $g'(y) = (1-y^2)^{-1} = 1+y^2+y^4+\cdots$, so the corresponding genus is the signature $L(M)$. As a cobordism invariant, this is characterized by its values on $\mathbb{C}P^{2k}$. Inspecting the cohomology ring (see the discussion of Chern classes) shows that this always equals 1. The formal group law is the addition formula for the hyperbolic tangent:

$$F_Q(x,y) = \frac{x+y}{1+xy} = x+y-xy(x+y)+\cdots.$$

Since $H^*(\mathbb{H}P^k, \mathbb{Z})$ is also a truncated polynomial ring on a 4-dimensional generator, similar considerations show that:

$$L(\mathbb{H}P^k) = \begin{cases} 0, & k = \text{odd} \\ 1, & k = \text{even} \end{cases}.$$

EXAMPLE 1.2. Consider the power series:

$$P(x) = \frac{x/2}{\sinh(x/2)}.$$

We obtain

$$g'(y) = \frac{1}{(1+(y/2)^2)^{1/2}}$$

and a formal group law given by the formula for $\sinh(x+y)$. This genus vanishes on all $\mathbb{H}P^k$, and geometrically it is related to the index $(\hat{A})$ of a certain elliptic differential operator, defined for Spin manifolds (see [53, Chap. 5 and appendix II]). Since we work over $\mathbb{Q}$, spin manifolds are sufficiently general.

EXAMPLE 1.3. Both previous examples are degenerate cases of the following universal elliptic genus whose logarithm is given by:

$$\log_\varphi(y) = \int_0^y \frac{dt}{(1-2\delta t^2+\varepsilon t^4)^{1/2}}.$$

If we write $\sqrt{R(t)}$ for the denominator, then:

$$F_\varphi(x,y) = \frac{x\sqrt{R(y)}+y\sqrt{R(x)}}{1-\varepsilon x^2 y^2}.$$

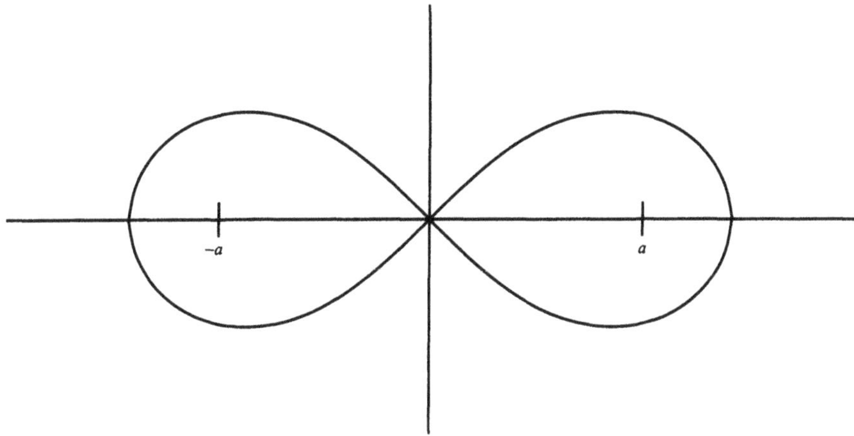

FIGURE 1.1. Lemniscate.

This addition formula arises for example in connection with the problem of doubling the arc length of the lemniscate (see Fig. 1.1) initiated by Fagnano in 1718, see [44]:

$$((x-a)^2+y^2)\cdot((x+a)^2+y^2)=a^4.$$

That is, the product of the distances to the points $(a,0)$ and $(-a,0)$ equals a^2.

The L-genus corresponds to ($\delta=\varepsilon=1$) and the $\hat{A}$-genus to ($\delta=-1/8$, $\varepsilon=0$); degeneracy means that the quartic equation has repeated roots, i.e., the discriminant $\Delta=\varepsilon^2(\delta^2-\varepsilon)=0$. In Chap. 4 we provide examples of nondegenerate elliptic genera associated with certain conjugacy classes of elements in the Mathieu group M_{24}. The preceding integral can be used to determine the value of any elliptic genus on the basic sequence $\{\mathbb{C}P^2,\mathbb{C}P^4,\dots\}$. For simplicity take $\varepsilon=1$, then classically the *Legendre polynomials* are defined by the expansion:

$$(1-2\delta t^2+t^4)^{-1/2}=P_0(\delta)+tP_1(\delta)+t^2P_2(\delta)+\cdots$$

This gives the general formula:

$$P_n(\delta)=\sum_{r=0}^{[n/2]}(-1)^r\frac{(2n-2r)!}{2^n r!(n-r)!(n-2r)!}\delta^{n-2r},$$

with initial values $P_0(\delta)=1$, $P_1(\delta)=\delta$, $P_2(\delta)=(3\delta^2-1)/2$, $P_3(\delta)=(5\delta^3-3\delta)/2,\dots$.

Alternatively the polynomial P_n arises as a solution of the ordinary differential equation $(1-t^2)y''-2ty'+n(n+1)y=0$. This in turn results from separating variables in Laplace's equation expressed in spherical polar coordinates (see [119] for more information).

EXAMPLE 1.4. We introduce the *universal formal group law*. Take $R = \Omega^*_{SO} \otimes \mathbb{Q}$ and let $i : \Omega^*_{SO} \to R$ be the natural map. Write

$$\underline{g}(y) = \sum_{n=0}^{\infty} \frac{1}{2n+1}[\mathbb{C}P^{2n}]y^{2n+1} \qquad \underline{F}(x,y) = x + y + \sum_{r,s} \underline{a}_{rs} x^r y^s$$

for the corresponding formal group law. Consider coefficients $\underline{a}_{rs}$ as equivalence classes of manifolds of dimension $2(r+s-1)$ (representatives are described later), and $\underline{a}_{rs} = 0$ if $r+s \equiv 0 \pmod 2$. The formula for $\underline{F}$ is an identity in the cobordism ring; hence it is respected by all homomorphisms $\varphi : \Omega^*_{SO} \otimes \mathbb{Q} \to R$. In particular coefficients a_{rs} in the power series for F_φ equal $\varphi(\underline{a}_{rs})$; this explains the use of the adjective universal.

1.3. Strong Multiplicativity

The discussion of Example 1.3 in Sec. 1.2 shows that any elliptic genus is determined by the values it takes on four- and eight-dimensional manifolds (indeed $\mathbb{C}P^2$ and $\mathbb{H}P^2$ suffice). This section shows that by taking a suitable basic sequence of polynomial generators, we can inductively build up values of φ starting at low dimensions. The approach is multiplicativity: By definition φ is compatible with Cartesian products, and we wish to extend this to certain types of locally trivial fibrations. Thus let:

$$\begin{array}{ccc} F & \longrightarrow & E \\ & & \big\downarrow \pi \\ & & X \end{array}$$

be a fibration of manifolds with structural group G. The genus φ is multiplicative for the fibration π if $\varphi(E) = \varphi(X)\varphi(F)$. This holds for $\varphi = L$ if G is connected and compact and for $\varphi = \hat{A}$ if F is a spin manifold. If the reader has not encountered $\hat{A}$ before, use the definition given by restricting the universal elliptic genus to the special values $\delta = -1/8$, $\varepsilon = 0$.

DEFINITION. The genus φ is strongly multiplicative if φ is multiplicative for spin fibrations with compact, connected structural group G.

THEOREM 1.3. *φ is strongly multiplicative if and only if φ is elliptic.*

We discuss only special cases of this: The full argument starts by proving that an elliptic genus is *rigid* with respect to a compact, connected group action [25, Theorem 3.7]. The theorem then implies that rigidity is equivalent to strong multiplicativity; we use a special case of this to complete the circle. More precisely we prove the following implications:

$$\varphi \text{ strongly multiplicative} \underset{(1)}{\Longrightarrow} \varphi \text{ elliptic}$$
$$\underset{(2)}{\Longrightarrow} \varphi \text{ vanishes on the class of } CP^{2n-1}\text{-fibrations.}$$

Recall that since $\mathbb{C}P^{2n-1}$ bounds, $\varphi(\mathbb{C}P^{2n-1})$ necessarily vanishes, proving multiplicativity in this case.

We start with the Milnor manifolds $H_{ij} \hookrightarrow \mathbb{C}P^i \times \mathbb{C}P^j, j \geqslant i$, defined as hypersurfaces of degree $(1,1)$ by the homogeneous equation $u_0v_0 + \cdots + u_iv_i = 0$. The H_{ij} is the total space of a $\mathbb{C}P^{j-1}$ fibration over $\mathbb{C}P^i$, and it can be given the structure of a smooth $2(i+j-1)$ manifold.

FACT 1. Evaluating the Pontrjagin numbers shows that classes $[H_{ij}]$ are a basic sequence of manifolds in $\Omega^*_{SO} \otimes \mathbb{Q}$ (see [86]).

FACT 2. If we write $\underline{H}(y_1,y_2) = y_1 + y_2 + \sum_{j+i\geqslant 2}[H_{ij}]y_1^iy_2^j$, then $\underline{H}(y_1,y_2) = \underline{g}'(y_1)\underline{g}'(y_2)\underline{F}(y_1,y_2)$. Here underscore refers to the universal formal group law (see Example 1.4).

Sketch of proof: With the same notation as before, if $u = g(y_1)$, $v = g(y_2)$, for any formal group law:

$$g(F(y_1,y_2)) = g(y_1) + g(y_2),$$

so that

$$F(y_1,y_2) = f(u+v) = \sum_{r,s} a_{rs}y_1^ry_2^s.$$

Now in the universal example, let u and v begin as cohomology classes in $H^2(\mathbb{C}P^i \times \mathbb{C}P^j, \mathbb{Z})$ obtained by pulling back polynomial generators for $H^*(\mathbb{C}P^i, \mathbb{Z})$ and $H^*(\mathbb{C}P^j, \mathbb{Z})$, respectively. *Homologically* classes u and v are represented by codimension 2 submanifolds, and the fact that H_{ij} has bidegree $(1,1)$ implies that it corresponds to $u+v$. More generally bidegree (a,b) describes a hypersurface dual to $au+bv$. Taking the ith and jth powers of u and v corresponds to self-intersecting $\mathbb{C}P^i$ and $\mathbb{C}P^j$ with themselves, so that geometrically we have the relation:

$$H_{ij} = \sum_{r,s} \underline{a}_{rs}\mathbb{C}P^{i-r}\mathbb{C}P^{j-s}.$$

Substitution in $\underline{H}(y_1,y_2)$ gives $\sum_{i,j}\sum_{r,s}\underline{a}_{rs}\mathbb{C}P^{i-r}\mathbb{C}P^{j-s}y_1^iy_2^j$, which on rearranging gives

$$\sum_{r,s\geqslant 0}(\underline{a}_{rs}y_1^ry_2^s)\left(\sum_{i\geqslant r}\mathbb{C}P^{i-r}y_1^{i-r}\right)\left(\sum_{j\geqslant s}\mathbb{C}P^{j-s}y_2^{j-s}\right)$$
$$=\underline{F}(y_1,y_2)g'(y_1)g'(y_2).\ \blacksquare$$

For more details, see [53, Chap. 3].

PROPOSITION 1.1. *The genus φ is elliptic if and only if φ vanishes on the Milnor manifolds $H_{3,2i}(i\geqslant 2)$.*

COROLLARY 1.1. *If the genus φ is strongly multiplicative, φ is elliptic.*

Proof of the Corollary: Since $H_{3,2i}$ is fibered by $\mathbb{C}P^{2i-1}$, $\varphi(H_{3,2i)}=0$. By Proposition 1.1 φ is elliptic. ■

Write $h_{ij}=\varphi(H_{ij})$ and replace variables y_1,y_2 by y,z, then:

$$h(y,z)=y+z+\sum_{i,j}h_{ij}y^iz^j=g'(y)g'(z)F(y,z).$$

If $r(y)=\sum_i h_{3,2i}y^{2i}$, what are the consequences of $r(y)=h_{3,2}h^2$?

Consider the Taylor expansion:

$$F(y,z)=y+F_z(y,0)z+\frac{1}{2}F_{zz}(y,0)z^2+\frac{1}{6}F_{zzz}(y,0)z^3+\cdots.$$

From the relation between F and its logarithm, we know that $g'(y)F_z(y,0)=1$, therefore:

$$h(y,z)=g'(z)\left(g'(y)y+z+\frac{1}{2}g'(y)F_{zz}(y,0)z^2\right)+\frac{1}{6}g'(y)F_{zzz}(y,0)z^3+\cdots,$$

with $g'(z)$ even. Sorting terms gives

$$r(y)=\frac{1}{6}g'(y)F_{zzz}(y,0).$$

Next we express $r(y)$ in terms of $b(y) = F_z(y,0) = 1/g'(y)$. Repeatedly differentiating $F(y,z) = g^{-1}(g(y)+g(z))$ with respect to z gives

$$F_z(y,z) = \frac{g'(z)}{g'(F(y,z))} = \frac{b(F(y,z))}{b(z)}$$

$$F_{zz}(y,z) = \frac{b'(F(y,z))b(F(y,z)) - b(F(y,z))b'(z)}{b(z)^2}.$$

Finally:

$$F_{zzz}(y,0) = b(y)[b''(y)b(y) + b'(y)^2 - b''(0)].$$

Note: $b'(0) = 0$.

Therefore $6r(y) = \{[b(y)^2]'' - b''(0)\}/2$; under our quadratic assumption on $r(y)$, this means that:

$$b(y)^2 = 1 - 2\delta y^2 + \varepsilon y^4 = R(y).$$

Conversely using this equation to define $b(y)$ forces the higher order terms in $r(y)$ to vanish.

Finally observe that $[b(y)^2]'' = -4\delta + 12\varepsilon y^2$, so that:

$$r(y) = h_{3,2}y^2 = \varepsilon y^2.$$

Therefore $\varepsilon = h_{3,2}$. The parameter δ is similarly determined by:

$$g'(y) = (1 - 2\delta y^2 + \varepsilon y^4)^{-1/2} = 1 + \delta y^2 \pmod{y^4},$$

so that from the definition of $\log_\varphi$, $\delta = \varphi(\mathbb{C}P^2)$. ■

As a corollary of the argument, an elliptic genus φ vanishes on all Milnor manifolds $H_{2i+1,j}$ for $j > i$. We have just shown this for $i = 1$, the more general case follows from the explicit elliptic formal group law. Thus:

$$F(y,z) = g^{-1}(g(y)+g(z)) = \frac{y\sqrt{R(z)} + z\sqrt{R(y)}}{1 - \varepsilon y^2 z^2},$$

and

$$h(y,z) = \frac{1}{\sqrt{R(y)}}\frac{1}{\sqrt{R(z)}}F(y,z)$$

$$= \left(\frac{y}{\sqrt{R(y)}} + \frac{z}{\sqrt{R(z)}}\right)(1 + \varepsilon y^2 z^2 + \varepsilon^2 y^4 z^4 + \cdots).$$

From this result by inspection it follows that $\varphi(H_{2i+2j+1,2j}) = 0$.

Now let us consider Implication 2; that is we want to prove that if φ is elliptic, it vanishes on all $\mathbb{C}P^{2n-1}$.

STEP 1: Determine the stable tangent bundle of the total space E of the fibration:

$$\begin{array}{ccc} \mathbb{C}P^{2n-1} & \longrightarrow & E \\ & & \downarrow \pi \\ & & B. \end{array}$$

Let ξ be the parent vector bundle over B with fiber $\mathbb{C}^{2n}$ and E' the associated S^{4n-1} bundle. The tangent bundle TE' is known to split as $(\pi')^! TM \oplus \eta$, where η is the bundle along the fibers. By equivariantly embedding a generic fiber in $\mathbb{C}^{2n}$ it is not difficult to see that $\eta \cong (\pi')^! \xi$. Replacing E' by E twists the fibers of η by $\overline{H}$, the conjugate bundle to the Hopf bundle, so that:

$$TE \cong \pi^! TM \oplus (\overline{H} \otimes \pi^! \xi).$$

STEP 2: The cohomological structure of E (compare Sec. 1.1).

As before write $t = c_1(\overline{H})$. A standard argument in algebraic topology shows that $H^*(E,\mathbb{C})$ is a free $H^*(B,\mathbb{C})$-module on basis $1, t, t^2, \ldots, t^{2n-1}$. The multiplicative structure is given by the unique relation:

$$t^{2n} + c_1(\xi) t^{2n-1} + \cdots c_{2n}(\xi) = 0. \qquad (1.1)$$

Indeed in some treatments, this relation from linear algebra is used to define Chern classes, furthermore if:

$$w = b_0 + b_1 t + \cdots b_{2n+1} t^{2n-1},$$

with $b_i \in H^*(B,\mathbb{C})$, then evaluation on fundamental classes gives $w[E] = b_{2n-1}[M]$. (A special case of this relates the top dimensional Chern class of ξ with its Euler number.) Let $\{K_k : k \geqslant 0\}$ be the multiplicative sequence corresponding to φ, then write

$$K(p.) = \sum_{k=0}^{\infty} K_n(p_1 \cdots p_k),$$

where p_i denotes the ith Pontrjagin class.

Then:

$$K(E) = \pi^* K(M) \cdot K(\overline{H} \otimes \pi^! \xi).$$

Given the preceding, it suffices to show that the coefficient of t^{2n-1} in the expansion of $K(\overline{H}\otimes\pi^!\xi)$ with respect to $\{1,t,\ldots,t^{2n-1}\}$ vanishes. Expressing the p_i as we may in terms of the formal variables $u_1,\ldots,u_{2n}$, we have

$$p_.(\pi^!\xi)=\prod_{k=1}^{2n}(1+u_k^2),$$

so that

$$K(\overline{H}\otimes\pi^!\xi)=\prod_{k=1}^{2n}\frac{t+u_k}{f(t+u_k)}. \tag{1.2}$$

Here f is the inverse to the logarithm $g=g_\varphi$ introduced in Sec. 1.2.

Equation (1.1) implies that $\prod_{k=1}^{2n}(t+u_k)=0$, and elementary manipulation shows that the product in (1.2) can be written as:

$$\psi(u_1,\ldots,u_{2n},t)+(\prod_{k=1}^{2n}(t+u_k))\alpha(u_1,\ldots,u_{2n},t),$$

where ψ is a polynomial of degree bounded by $2n-1$ in t. More explicitly write $Q(u)=Q^*(u^2)=u/f(u)$. Then for each i:

$$\psi(u_1,\ldots u_{2n},-u_i)=\prod_{k\neq i}Q(u_k-u_i).$$

Note: $Q(0)=1$. Therefore we must have

$$\psi(u_1,\ldots,u_{2n},t)=\sum_{i=1}^{2n}\prod_{k\neq i}\left[Q(u_k-u_i)\frac{u_k+t}{u_k-u_i}\right],$$

where terms in this equation lie in the field of fractions for the ring of formal symmetric power series in variables $u_1\cdots u_{2n}$. The coefficient of t^{2n-1} equals

$$\sum_{i=1}^{2n}\prod_{k\neq i}[f(u_k-u_i)]^{-1}.$$

STEP 3: Implications of ellipticity. In the remainder of the argument we use the most elementary properties of doubly periodic functions, see, for example, [73].

The elliptic integral defining $g=\log_\varphi$ implies that $\mu(u)=1/f(u)$ is meromorphic on a certain torus $(\mathbb{C}/2L)^{2n}$. Here the basic parallelogram for L is determined by points ω_1,ω_2, and we write $\omega=\omega_1+\omega_2$. To prove that the coefficient of t^{2n-1} above is zero, we must show that

$$\sum_{i=1}^{2n}\prod_{k\neq i}\mu(u_k-u_i)$$

vanishes for all $2n$-tuples lying in some nonempty open subset U of the torus. Take U to contain all points $(u_1,\ldots,u_{2n})$ such that $u_i \not\equiv u_j$ (mod the lattice) if $i \neq j$. The function $\lambda(u) = \prod_{k=1}^{2n} \mu(u_k - u)$ is still $2L$ elliptic, and by choice of U has $2(2n)$ distinct poles:

$$u_1,\ldots,u_{2n},u_1+\omega,\ldots u_{2n}+\omega,$$

with residues $\prod_{k\neq i} \mu(u_k - u_i)$ at points $u_1,\ldots,u_{2n}$.

This follows since the residue of $\mu(u)$ at $u=0$ equals 1. Similarly the residue at $u=\omega$ is -1, so that the residue of $\lambda(u)$ at $u_i+\omega$ equals

$$-\prod_{k\neq i} \mu(u_k - u_i - \omega) = \prod_{k\neq i} \mu(u_k - u_i),$$

since

$$\mu(u+\omega) = -\mu(u)$$

and there are an odd number of factors in the product. However ellipticity implies that the sum of *all* the residues must vanish, as required. ∎

1.4. Notes

Chapter 1 is based on [53, Chaps. 1 and 3], although in proving that a strongly multiplicative genus is elliptic, we follow the original treatment in [90]. As lucidly explained by G. Segal, strong multiplicativity should be seen as a byproduct of rigidity. This term can be explained as follows: Using the Chern character, φ can be expressed in terms of Λ_φ, a stable exponential characteristic class taking values in $K(\cdot)\otimes\mathbb{C}$. We have the formula $\varphi(M) = \pi_!^M(\Lambda_\varphi TM)$, where $\pi_!^M : K(M)\otimes\mathbb{C} \to \mathbb{C}$ is the Gysin map associated with the fibration of M over a point. All this makes sense equivariantly, and in the presence of a compact G-action, $\varphi(M)$ should be considered as a class function. To say that $\varphi(M) = \varphi_G(M)$ is *rigid* means that if G is *connected* as well as compact, then $\varphi_G(M)$ is constant as a function of G. We emphasize the connectedness, since in Chap. 4 we will define a different kind of equivariant genus for the discrete group M_{24}. For more on the important concept of rigidity, see [25].

If the reader wants a rapid course in the necessary algebraic topology, Ref. [86] on characteristic classes is strongly recommended. For a quick introduction to Chern and Stiefel–Whitney classes, also see in the book on fiber bundles by D. Husemoller [60, Chap. 16]. Specific references include [60, Subsection 16.5] on the splitting principle and [60, Subsection 16.2.5] on the significance of Eq.

(1.1). Reference [86] also contains a good introduction to cobordism and the role of characteristic numbers in detecting cobordism classes.

Multiplicative sequences were introduced by F. Hirzebruch, who paid special attention to the signature (L) and Todd genus (for stably complex manifolds). The $\hat{A}$-genus, which we introduced as a limiting example of the universal genus, is classically defined in terms of the index of an elliptic (Dirac) operator. For this see the readable account by P. Baum in [53, appendix II].

2

Cohomology Theory $Ell^*(X)$

2.1. Motivation

The ring Ω^*_{SO} introduced in Chap. 1 can be regarded as coefficients of a generalized (co)homology theory. As we explain in more detail in Sec. 2.3, the bordism group $\Omega^{SO}_m(X)$ for a suitable CW-complex X consists of equivalence classes of maps $f : M^m \to X$, where $f_0 \sim f_1$ if these are obtained by restricting some map $F : W^{m+1} \to X$ to $M_0 \cup (-M_1) = \partial W$. In the 1960s [34] showed that K-theories associated with various classes of bundles are actually quotients of cobordism, whose complex variant is universal from the point of view of formal group laws (compare examples in Chap. 1). Implicit in their work is the isomorphism:

$$\Omega^*_{\text{Spin}}(X) \underset{\Omega^*_{\text{Spin}}}{\otimes} KO^*(\text{point}) \cong KO^*(X),$$

where $KO^*(pt)$ is made into an Ω^*_{Spin}-module by modifying the definition of the $\hat{A}$-genus to allow for copies of $\mathbb{Z}/2$ in dimensions $\equiv 1, 2 \pmod 8$. This has target KO^* (point) rather than the integers, so we map S^1 to s, the class of the Kummer surface to k, and M^8 with $\hat{A}(M^8) = 1$, $sig(M^8) = 0$ to b, where s, k, and b are generators in real K-theory. The main ingredients in the Conner–Floyd argument are (1) the existence of a KO^*-Thom class for spin bundles (see [15]) and (2) the simple polynomial structure of both KO^* and Ω^*_{Spin} on complex projective space (at least away from the prime 2).

The result quoted strongly suggests that elliptic cohomology, defined in Sec. 2.4, should also be obtained from a suitable class of vector bundles. For certain spaces we attempt to lay the foundations of such a construction in Chap. 8.

Additional motivation for regarding the ring where the universal elliptic genus φ takes its values as potentially interesting from the point of view of coefficients

follows. We already saw that φ defines a map:

$$\varphi : \Omega^*_{SO} \otimes \mathbb{Z}\left[\frac{1}{2}\right] \to \mathbb{Z}\left[\frac{1}{2}\right][\delta, \varepsilon],$$

where it is permissible to replace $\mathbb{Q}$ by $\mathbb{Z}[1/2]$ on the grounds that the only torsion in cobordism is 2-torsion. For reasons that become clear, we give the generators δ and ε on the right negative degrees 4 and 8, respectively. If the integral defining $\log_\varphi$ is such that $\Delta = \varepsilon^2(\delta^2 - \varepsilon)$ (of degree 24) is nonzero, then it gives rise to an elliptic curve with equation:

$$y^2 = 4x^3 - g_2 x - g_3 = 4(x - e_1)(x - e_2)(x - e_3),$$

where $\delta = -3/2e_1$, $\varepsilon = (e_1 - e_2)(e_1 - e_3)$, and $e_1 + e_2 + e_3 = 0$.

This correspondence between quartic and cubic curves is classical. If we start with the given cubic equation and $\mathfrak{P}$ denotes the parametrizing function, then the associated lattice L has basis given by

$$e_1 = \mathfrak{P}\left(\frac{\omega_1}{2}\right), \qquad e_2 = \mathfrak{P}\left(\frac{\omega_2}{2}\right),$$

in addition:

$$e_3 = \mathfrak{P}\left(\frac{\omega_1 + \omega_2}{2}\right).$$

Now let $f(z) = [\mathfrak{P}(z) - e_1]^{-1/2}$, and consider the curve in $\mathbb{C}P^2$ parametrized by $[\xi = f(z),\ \eta = f'(z)]$. With $(x,y) = (\mathfrak{P}(z), \mathfrak{P}'(z))$ an easy calculation shows that $(x,y) = ((\xi^{-2} + e_1), -2\eta\xi^{-3})$.

Therefore substituting into the cubic equation, we have

$$4\eta^2\xi^{-6} = 4\xi^{-2}(\xi^{-2} + e_1 - e_2)(\xi^{-2} + e_1 - e_3),$$

or

$$\begin{aligned} \eta^2 &= (1 + \xi^2)(e_1 - e_2)[1 + \xi^2(e_1 - e_3)] \\ &= 1 + (3e_1 - e_1 - e_2 - e_3)\xi^2 + (e_1 - e_2)(e_1 - e_3)\xi^4, \end{aligned}$$

justifying the defining equations for δ and ε. Note: The lack of a square term in the cubic implies that $e_1 + e_2 + e_3 = 0$. The function f that parametrizes the quartic curve has lattice $\widetilde{L}$ of index 2 in L, and the projection $\mathbb{C}/\widetilde{L} \to \mathbb{C}/L$ is an unbranched double covering (see Fig. 2.1).

The odd function f has zeros of order 1 at all L-lattice points and poles of order 1 at $(\omega_1/2) + L$. In our notation this distinguishes ω_2 as a zero, normalized as $\omega_2 = 2\pi i\tau$ with τ in the upper half-plane H.

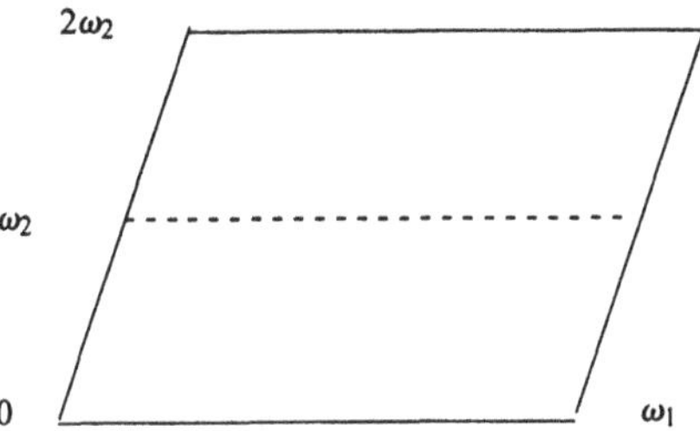

FIGURE 2.1. Fundamental parallelogram.

Returning to the integral $\log_\varphi$, we see that replacing (δ, ε) by $(\lambda^2\delta, \lambda^4\varepsilon)$ changes the lattice L to $\lambda^{-1}L$ and the genus $\varphi(M)$ to $\lambda^{(\dim M)/2}\varphi_M$. Hence as a function of τ, $\varphi(M)$ is a modular form of weight $(\dim M)/2$ with invariance subgroup $\Gamma_\circ(2) \subset PSL_2(\mathbb{Z})$ preserving ω_2. [Algebraically $\Gamma_\circ(2)$ is defined by 2 x 2 matrices $\left(\begin{smallmatrix} a & b \\ c & d \end{smallmatrix}\right)$ with $c \equiv 0 \pmod 2$. The degenerate cases L and $\hat{A}$ correspond to cusps of $H/\Gamma_0(2)$ at points 0 and $i\infty$.] Most importantly we can identify the future ring of coefficients $\mathbb{Z}[1/2][\delta, \varepsilon]$ with the ring of all modular forms of even weight and group $\Gamma_0(2)$, whose Fourier expansions at the cusps have coefficients belonging to $\mathbb{Z}[1/2]$ (see [53, appendix 1], for the role played by δ and ε in this result).

2.2. Complex-Oriented Cohomology Theories

We start with a generalized cohomology theory $h^*(\cdot)$ with products where $1/2$ belongs to the zero summand of the ring of coefficients $h^* = h^*$ (point). Such a theory obeys the Eilenberg–Steenrod axioms (see any good text on algebraic topology). for example there is a long exact Mayer–Vietoris sequence associated with the space $X = A \cup B$, $C = A \cap B$, and homotopic maps induce the same homomorphisms of cohomology groups. Additional structure is associated with an *orientation* $t \in h^2(\mathbb{C}P(\infty))$, which is such that:

- t maps to $-t$ under complex conjugation,
- t restricts to the canonical generator of $h^2(S^2)$,

(obtained for example by suspension). The existence of such an element allows us to mimic constructions in ordinary cohomology and in particular to obtain

1. Chern classes $c_i^{(h)}(E) \in h^{2i}(X)$ for a complex bundle E over the base space X,

2. A Thom isomorphism $h^i(X) \cong \widetilde{h}^{i+n}(E^+)$ for each oriented real n-dimensional vector bundle E over X,
3. Gysin maps $f_* : h^i(M) \to h^{i-n}(X)$ for each fibration $f : E \to X$, fibered by an oriented n-manifold,

where E^+ denotes the Thom space of E, obtained by coning off the boundary of an associated unit disc bundle and the reduced group $\widetilde{h}^i(X) = h^i(X,\text{point})$. As in the classical case Case 1 follows from the polynomial structure of $h^*(\mathbb{C}P^{(\infty)})$, Case 2 is first proved for a product (where the compactification of a typical fiber is S^n), then extended to the general case by a Mayer–Vietoris argument; and Case 3 uses Poincaré duality to reverse the direction of the induced map f^*. We can now define

4. An h^*-valued genus φ_h with $\varphi_h(M) = f_*(1) \in h^{-\dim(M)}$, where the constant map $f : M \to$ point is regarded as a fibration.
5. A graded odd formal group law F over h^* given by:

$$F(c_1^{(h)}(L), c_1^{(h)}(M)) = \sum_{i,j} a_{ij} c_1^{(h)i}(L) c_1^{(h)j}(M) = c_1^h(L \otimes M).$$

That is, we take the Chern class of the tensor product of two line bundles. To say that F is odd and graded means that $a_{ij} \in h^{-2(i+j-1)}$ and $F(-x,-y) = -F(x,y)$. Assuming that we know the universal property of the formal group law in cobordism ($h^* = \Omega^*_{SO} \otimes \mathbb{Z}[1/2]$), Cases 4 and 5 are equivalent (compare the discussion in Chapter 1).

There is a close connection, explained for example in [96, Chap. 4], between oriented cohomology theories over $\mathbb{Z}[1/2]$ and anticommutative graded rings R equipped with a formal group law F. Given such a pair (R,F) we first use the universal property of $\Omega^*_{SO} \otimes \mathbb{Z}[1/2]$ to define a homomorphism (genus) φ : $\Omega^*_{SO} \otimes \mathbb{Z}[1/2] \to R$, hence a left module structure on R, and then set

$$h^*(X) = \left(\Omega^*_{SO}(X) \otimes \mathbb{Z}\left[\frac{1}{2}\right]\right) \underset{\varphi}{\otimes} R.$$

Provided certain flatness conditions are satisfied, and we see later that this is the case for suitable localizations of $\mathbb{Z}[1/2][\delta,\varepsilon] = R$, such quotients of cobordism are oriented cohomology theories. We can indeed state the flatness conditions locally for each prime p in terms of specific cobordism classes v_n. These were introduced in a purely algebraic way by E. Brown and F. Peterson [27] in the course of their description of the p localization of cobordism in terms of building blocks isomorphic to $\mathbb{Z}_{(p)}[v_1, v_2, \dots]$ with $\deg(v_i) = 2(p^i - 1)$. At root these classes v_i are associated with monomials in (mod p)-cohomology operations describing a $\mathbb{Z}_{(p)}$-basis for the Steenrod algebra $\mathfrak{A}_p$ (modulo the obvious Brockstein map). Note:

the ring is much smaller than $\Omega^*_{SO} \otimes \mathbb{Z}[1/2]$, which has a polynomial generator in *every* even dimension. The ideal I_n generated by $(p = v_0, v_1, \ldots, v_{n-1})$ is shown to be prime and hence *regular* in the sense that v_i is not a zero-divisor modulo the ideal generated by $v_j (0 \leqslant j \leqslant i-1)$ (see Appendix A for more details and an alternative approach to the generators v_i).

The Landweber exact functor theorem (Theorem A.2) with a sketch proof in [97] now states that:

$$h^*(X) = \left(\Omega^*_{SO}(X) \otimes \mathbb{Z}\left[\frac{1}{2}\right]\right) \underset{\varphi}{\otimes} R$$

is a cohomology theory if for each prime p, images of the v_i under φ form a regular sequence in R. Finally we say that the (p-local) theory h^* is v_n-periodic if $\varphi(v_n) = u_n$ is invertible in R. All these notions play an important part in Chap. 3, but for the moment we make a geometric digression in the hope of reviving the nonspecialist's flagging interest.

2.3. Baas–Sullivan Construction

We return to the bordism theory $\Omega^{SO}_*(X)$ mentioned in Sec. 2.1. These groups form a *homology* theory in the sense that all Eilenberg–Steenrod axioms except the dimension axiom are satisfied. Recall that $\Omega^{SO}_n(X)$ consists of equivalence classes of maps $f : M^n \to X$, where M^n is an oriented smooth manifold. We define relative groups $\Omega^{SO}_*(X,A)$ by means of maps of pairs $(M^n, \partial M^{n-1}) \xrightarrow{f} (X,A)$, and we define boundary maps $\partial : \Omega^{SO}_n(X,A) \to \Omega^{SO}_{n-1}(A)$ by $\partial[M^n, f] = [\partial M^{n-1}, f|\partial M^{n-1}]$. Checking functoriality and the dependence of the induced map g_* only on the homotopy class of $g : (X,A) \to (Y,B)$ is straightforward. The remaining axioms exploit the following:

FACT. If M^n is a closed manifold containing V^n as a submanifold with boundary and $f : M^n \to X$ is such that $f(M^n \setminus \mathrm{Int}(V^n)) \subseteq A$, then $[M^n, f] = [V^n, f|V^n]$ in $\Omega^{SO}_n(X,A)$.

Exactness Property: The following sequence of abelian groups is exact:

$$\Omega^{SO}_n(A) \underset{i^*}{\longrightarrow} \Omega^{SO}_n(X) \underset{j_*}{\longrightarrow} \Omega^{SO}_n(X,A) \underset{\partial}{\longrightarrow} \Omega^{SO}_{n-1}(A).$$

Most of the checks are trivial; the preceding fact is needed to show $j_* i_* = 0$ and $\ker \partial \subseteq \mathrm{Im} j_*$. The first of these is straightforward; for the second observe that if the pair $f : (V, \partial V) \to (X,A)$ is such that $f|\partial V$ extends to a map $g : U \to A$, the manifold $M = V \cup_{\partial V} U$ and the map $f \cup g$ satisfy the condition.

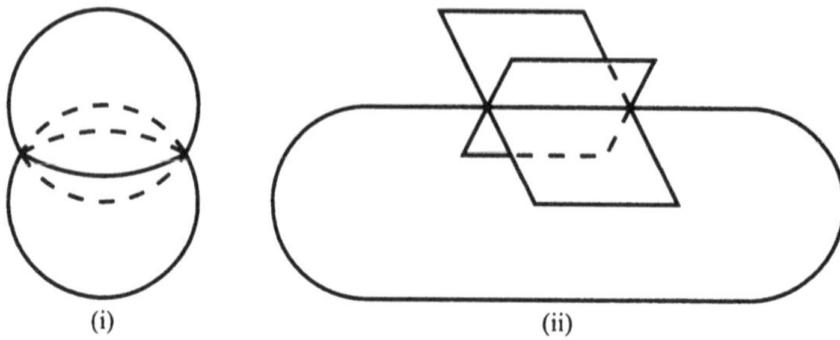

FIGURE 2.2. Baas–Sullivan construction.

Excision Property: If $\overline{U} \subset A$, then $i_* : \Omega_n^{SO}(X \backslash U, A \backslash U) \cong \Omega_n^{SO}(X, A)$. To show that i_* is a surjective map, let $[B^n, f] \in \Omega_n^{SO}(X, A)$, define $P = f^{-1}(X \backslash A)$, $Q = f^{-1}(\overline{U})$ and separate these closed subsets with a manifold B_1^n (with boundary). Another application of the preceding fact shows that:

$$i_*[B_1^n, f|B_1^n] = [B^n, f].$$

Injectivity is similar.

(Co)bordism has its origins in the attempt to represent the ordinary homology classes of a smooth manifold in terms of embedded submanifolds. In general this is not possible, but the situation becomes a little easier if we allow manifolds with singularities.

DEFINITION. A closed n-dimensional manifold with singularity of type (P_1), dimension $P_1 = k_1 < n$, is a space:

$$W = A \cup_{A(1) \times P_1} (A(1) \times Cone(P_1)),$$

where $\dim(A(1)) = n - k_1 - 1, \partial A = A(1) \times P_1$ and A and $A(1) \times Cone(P_1)$ are glued together along ∂A. The space W bounds if there exists V such that:

$$V = B \cup_{B(1) \times P_1} (B(1) \times Cone(P_1))$$

with $\partial B = A_{\cup} B(1) \times P(1)$ and $\partial B(1) = A(1)$.

As an example consider the intersecting copies of S^2 in Fig. 2.2.

A neighborhood of the singularity set takes the form $S^1 \times C$(4 points), and on its removal, the complement is $\mathring{D}^2 \times$ (4 points). Hence W can be written as $[D^2 \times$ (4 points)$] \cup [S^1 \times Cone$(4 points)$]$.

It is possible to iterate this construction, introducing singularities of the type $\{P_0 = \text{point}, P_1, P_2, \ldots, P_m\}$, and denoting the associated bordism theory by $\Omega(S_n)_*(X,A)$. Verification of the Eilenberg–Steenrod axioms proceeds much as for the nonsingular theory $\Omega(S_0)_*$. The only problems occur with exactness of the long homology sequence and excision. The former is handled by keeping a careful eye on the singular set, the latter inductively. Thus there is another long exact sequence linking two copies of $\Omega(S_n)_*$ and one of $\Omega(S_{n+1})_*$, to which the standard 5-Lemma in homological algebra applies. The induction starts with the preceding sketch of a proof of excision for $\Omega(S_0)_*$; for all this, see [17].

Manifolds $P_1, P_2, \ldots$ represent classes in the coefficient ring Ω^*_{SO}, and it is a natural question to compare the theory dual to $\Omega(S_n)_*$ with $\Omega^*_{SO}(\cdot) \otimes R$, where φ is the projection map onto $\Omega^*_{SO}/\langle P_i : i \geqslant 0\rangle$. As we may expect, we obtain a good correspondence whenever the sequence $\{P_i : i \geqslant 0\}$ is regular. For example in a rather crude way, we can recover the Brown–Peterson theory by killing the kernel of the map $\Omega^{SO}_* \to BP_*$, then localizing at p. We sketch the main steps of the argument partly because it enables us to construct elliptic cohomology without saying too much about elliptic curves and also because of its relevance to other theories of elliptic type discussed later.

2.4. Construction of $Ell_*(X)$ and $Ell^*(X)$ away from the Prime 2

Let $R = \mathbb{Z}[1/2][\delta, \varepsilon]$ be the ring introduced in Sec. 2.1, and consider the diagram of ring homomorphisms:

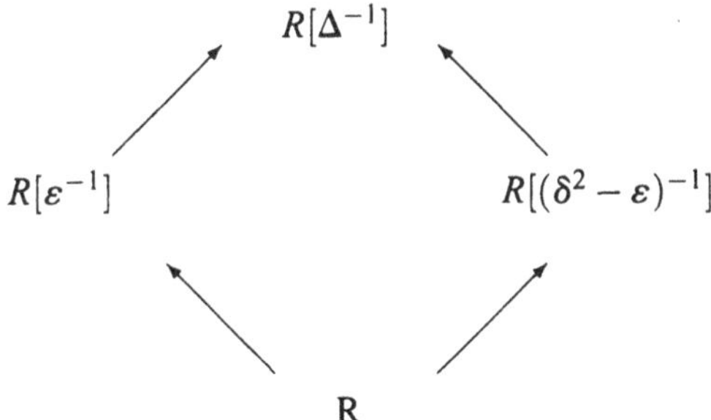

THEOREM 2.1. *There are homology theories having each of these rings as coefficients. Each theory is multiplicative, and the corresponding cohomology theory is complex-oriented. In all three cases the logarithm of the formal group law is that of the universal elliptic genus.*

Proof: We need the Baas–Sullivan construction, the theorem that $\Omega_*^{SO} \otimes \mathbb{Z}[1/2]$ can be generated by Milnor manifolds $H_{i,j}$, and Ochanine's characterization of the ideal killed by the universal elliptic genus (see Theorem 1.3).

Write $\Omega_*^{SO} \otimes \mathbb{Z}[1/2] = \mathbb{Z}[1/2][x_4, x_8, x_{12}, \dots]$, where $x_4 = [\mathbb{C}P^2], x_8 = [H_{3,2}]$, and the remaining elements $x_{4k} (k \geqslant 3)$ generate the ideal killed by the genus φ. This is possible, since each x_{4k} can be chosen as the total space of a $\mathbb{C}P^{2m-1}$-fibration. With $\{x_{4k} : k \geqslant 3\}$ as the singularity set and x_4 and x_8 mapped to δ and ε, respectively, we obtain a theory whose coefficients are $\mathbb{Z}[1/2][\delta, \varepsilon]$ modulo questions of regularity. In this case these are trivial, since x_{4k} is indecomposable in a polynomial ring. Since 2 has been inverted, obstructions to the existence of a product vanish (see [87]), i.e., the homology theory obtained with coefficients R is multiplicative. The statement about the formal group law follows, since to give this is equivalent to giving a homomorphism from the universal example into R (or into one of its extensions). ■

The argument just given is topological, and although Ochanine's theorem uses some properties of elliptic functions, their role is very much hidden. His theorem does have the advantage of applying to situations where we may not have good arithmetic information, as we see when discussing [67] in Chap. 7. However once we know the existence of a universal elliptic genus φ, we can form the quotient theory $\Omega_*^{SO}(X) \underset{\varphi}{\otimes} R$ directly, then check the regularity of the sequence $\{p, u_1, u_2, \dots\}$ for each prime $p \geqslant 3$. (Here u_j denotes the image of the generator v_j in BP_* under the localized homomorphism φ.) At this point we also replace the ring R by one of $R[\Delta^{-1}], R[\varepsilon^{-1}]$ or $R[(\delta^2 - \varepsilon)^{-1}]$, which introduces periodicity into the cohomology theory dual to the homology theory just constructed. Hence we need check only regularity for the sequence $\{p, u_1, u_2\}$, which results from the following assertions:

1. Multiplication by p is injective.
2. $u_1 \equiv P_{(p-1)/2}(\delta, 1)$ modulo p; since the Legendre polynomial takes the value 1 at 1, $u_1 \not\equiv 0 \pmod{p}$. This follows from

LEMMA 2.1. *In the power series ring $\mathbb{F}_p[[x]]$ we have*

$$(1+x)^{-1/2} = \lim_{l \to \infty} [(1+x)^{(p-1)/2}]^{1+p+\cdots+p^{l-1}}$$

Proof: Completion is with respect to the (x)-adic topology. The assertion is equivalent to showing that:

$$\lim_{l \to \infty} [(1+x)^{1/2}]^{(p-1)(1+p+\cdots+p^{l-1})+1} = 1,$$

i.e., $\lim_{l\to\infty}[(1+x)^{1/2}]^{p^l} = 1$.

This is obvious, since over $\mathbb{F}_p$, $(1+x)^{p^l} = 1+x^{p^l}$ and high powers of x tend to 0 (x)-adically. ■

COROLLARY 2.1. *If P_n denotes the nth Legendre polynomial, as defined in Chap. 1, then:*

$$\sum_{n=0}^{p-1} P_n(x)t^{2n} \equiv (1-2xt^2+t^4)^{(p-1)/2} \pmod{p}.$$

Proof: This follows by replacing $(1+x)$ by $(1-2xt^2+t^4)$ in the preceding argument. Now compare coefficients to determine $u_1 \pmod{p}$. ■

Rather harder to prove [70, Sec. 6] is Assertion 3:
Modulo (p, u_1)

$$u_2 \equiv (-1)^{(p-1)/2}\varepsilon^{(p^2-1)/4} \equiv (-1)^{(p-1)/2}(\delta^2-\varepsilon)^{(p^2-1)/4}$$

in the ring $\mathbb{Z}[1/2][\delta,\varepsilon]$. Hence on inverting any one of ε, $(\delta^2-\varepsilon)$ or Δ, u_2 becomes a unit, and the sequence $\{p, u_1, u_2\}$ is regular.

We reduce Assertion 3 to a question about the formal group law on a Weierstrass elliptic curve by first using the dictionary for passing between cubics and quartics, given in Sec. 2.1, then proving that the two formal groups' laws are strictly isomorphic (see [96] for the significance of this). Since we already checked that $\{p, u_1\}$ is regular, we can restrict attention to the so-called supersingular elliptic curves, for which we have the following proposition.

PROPOSITION 2.1. *Let E be a supersingular curve over a field of characteristic $p \geqslant 5$, so that*

$$[p](z) = F(z, [p-1]z) = u_2 z^{p^2} + \cdots$$

($u_2 \neq 0$, $z = -x/y$ equals the standard uniformizing parameter, and F is the formal group law). Then $u_2 = (-1)^{(p-1)/2}\Delta^{(p^2-1)/12}$.

P. Landweber [69] attributes this result to B. Gross [51]. Note: $p \geqslant 5$ is not a handicap, since the invertibility of u_2 can be checked by hand for $p = 3$. Indeed with $\varepsilon = 1$, and in terms of the variable x (compare Lemma 2.1) $u_1 = x$ and $u_2 = (9x^4 - 10x^2 + 1)/8$.

The proof of Proposition 2.1 relies heavily on numerical results from the theory of elliptic curves (see [66, 70]). Again we try to give the main ideas behind the argument.

Having eliminated special cases, we show that the curve in question can be assumed to be defined over $\mathbb{F}_{p^2}$ and multiplication by p in the formal group law corresponds to the same operation on the curve E. Then the Frobenius map of degree p^2 satisfies $Fr = [\pm p] : E \to E$. Turning to the formal group, we notice that the induced Frobenius map sends z to z^{p^2}, we have

$$u_2 z^{p^2} + \text{higher terms} = [\pm 1] z^{p^2},$$

implying that $u_2 = \pm 1$. This is enough for regularity, but to decide which elements in $\mathbb{Z}[1/2][\delta, \varepsilon]$ are suitable to invert, we must compare this calculation with the one for Δ. To do this we assume that $u_2 = (-1/p)$ (Legendre symbol), circumventing this restriction later by introducing a twist of E. Proposition 2.1 follows if we show that $\Delta^{(p^2-1)/12} = 1$. This proceeds in two stages:

1. The relations $Fr = [(-1/p)p] = 1 \pmod 4$ show that the 4-torsion subgroup of E is defined over $\mathbb{F}_{p^2}$. A classical formula then says that Δ is a fourth power in $\mathbb{F}_{p^2}$, so $\Delta^{(p^2-1)/4} = 1$.

2. Similar considerations with the 3-division points of E show that $\Delta^{(p^2-1)/3} = 1$.

We summarize what we have constructed as follows.

There exists a complex-oriented multiplicative cohomology theory, $Ell^*(X)$, that is v_2-periodic for all odd primes and has coefficients $\mathbb{Z}[1/2][\delta, \varepsilon, \varepsilon^{-1}]$, where $\deg(\delta) = -4, \deg(\varepsilon) = -8$. The coefficient ring can be regarded as a localization of the ring of $\Gamma_0(2)$-invariant modular forms of even weight with the Fourier coefficients of the expansions about the two cusps belonging to $\mathbb{Z}[1/2]$.

2.5. Notes

There are many good references for generalized cohomology theories; the treatment given here perhaps owes most to various lectures given over the years by A. Dold (see for example [41]). But anyone seriously interested in the foundations of the subject should look at the collection of various lecture courses in Chicago by J. F. Adams [1]. These include far more material than required just to construct elliptic cohomology: Part 1 deals with (complex) cobordism as the master theory, Part 2 with formal group laws including the local version for Brown–Peterson theory (BP). Part 3 is more foundational, and it should perhaps be looked at first.

Another older reference written before the development of the language of spectra, but with good geometric motivation is [43].

On the arithmetic side [66] is an excellent reference for the theory of elliptic curves, the most relevant chapters being 3 ("Cubic Curves"), 6 ("Complex Points") and 8 ("Modular Forms"). There is also an introduction to modular forms in [53, appendix I], which explains, for example, the pivotal role of the discriminant Δ in providing periodicity for both $Ell^*(X)$ and the ring of modular forms.

3

Work of M. Hopkins, N. Kuhn, and D. Ravenel

3.1. Bundles over the Classifying Space BG

Throughout Chap. 3 G denotes a finite group, topologized where necessary with the discrete topology. If X is a suitable space, for example a CW-complex with finitely many cells in each dimension, principal G-bundles over X are classified up to isomorphism by homotopy classes of maps from X into the so-called classifying space BG. This is a space of the same kind as X, characterized by the properties (1) $\pi_0 BG = \{0\}$, i.e., BG is connected, (2) $\pi_1 BG = G$, and (3) $\pi_i BG = \{0\}, i \geqslant 2$. It is usual to denote the universal covering space by EG; this has the structure of a principal G-bundle, and it is universal in the sense that any such bundle ξ over X is of the form $f^! EG$, which is induced by some map $f : X \to BG$. For all this, see again [60]. There is a natural map:

$$\alpha : R(G) \to K(BG)$$

from the complex representation ring of G to the Grothendieck group of complex vector bundles over the space BG. This map sends the class of a $\mathbb{C}G$-module M to the class of the complex vector bundle associated to EG by the G-action on M. Thus if we allow G to act on EG from the right and on M from the left, the total space of the bundle $\alpha(M)$ [fibered by $\mathbb{C}^n, n = \dim(M)$] is given by $EG \underset{G}{\times} M$. Here we identify (e,m) with (eg,gm), then give the resulting space the quotient topology; the local triviality of EG ensures that of $\alpha(M)$. We refer to α as the flat-bundle homomorphism — [Flatness refers to the fact that the discreteness of the topology on G allows us to construct a flat connection in the bundle $\alpha(M)$]. A remarkable paper [13] showed that the image of α is dense in $K(BG)$. To make this precise, we topologize $R(G)$ by defining powers of the augmentation ideal

$I(G) = \ker\{R(G) \underset{\dim}{\longrightarrow} \mathbb{Z}\}$ to be a basic neighborhood system at the identity. The completion $R(G)_I^\wedge$ then equals $\overset{\lim}{\underset{n}{\leftarrow}} (R(G)/I^n(G))$. For the completed flat bundle homomorphism $\alpha^\wedge$ we have Theorem 3.1.

THEOREM 3.1. *$\alpha^\wedge : R(G)_I^\wedge \to K(BG)$ is an isomorphism.*

REMARKS. The topology on $K(BG)$ is defined by taking the inverse limit over finite skeletons, and $\alpha^\wedge$ is compatible with both algebraic and topological structure. Although a more elegant proof of Theorem 3.1 was subsequently given in [12] by embedding G in some unitary group, the original proof is the basis of methods in this chapter. In this proof: (1) $\alpha^\wedge$ is an isomorphism for cyclic groups $C_p \cong \mathbb{Z}/p$; the clue to this is the fact that BC_p can be taken as the infinite dimensional lens space $\underset{\longleftarrow}{\lim}(S^{2n-1}/C_p)$, where C_p is allowed to act on S^{2n-1} through a sum of one-dimensional representations. This result (2) is extended inductively to solvable groups and in particular to direct products $C_r \times P$, where P is a p-group $(p,r) = 1$ (elementary group); (3) all groups are included by using a completed version of the Brauer induction.

The case of a p-group P is rather special in that:

- the natural map $R(P) \longrightarrow R(P)_I^\wedge$ is injective.
- I-adic completion may be replaced by p-adic completion.

This means for example that a class in $K(BC_p)$ is represented by $\sum_{r=0}^{p-1} a_r \xi^r$, where a_r belongs to the p-adic integers $\mathbb{Z}_p$ and ξ is the representation that maps a fixed chosen generator of C_p to $e^{2\pi i/p}$. This is actually the first step in the [2] characterization of the image of $R(G)$ (uncompleted) in $K(BG)$.

Thus for a p-group Theorem 3.1 says that $R(P)_p^\wedge \cong K(BP)$. We wish to extend this result to other cohomology theories of bundle type. Another hint that this may be possible is provided by the isomorphism between the representation ring $R(G)$ (defined in terms of $\mathbb{C}G$-modules) and its character ring (defined in terms of class functions).

3.2. General Character Theory

In Chap. 5 we discuss Mackey functors, i.e., functors from groups to rings with induction and restriction maps satisfying Frobenius reciprocity and having a double-coset formula. A first example of such a functor is $\{H \rightsquigarrow R(H) : H \subseteq G\}$. We also have the character map:

$$\begin{aligned} \chi : R(G) &\longrightarrow Cl(G) \\ M &\longmapsto \{\mathrm{trace}(g : M \to M) : g \in G\}, \end{aligned}$$

where $Cl(G)$ denotes the space of complex-valued functions that are constant on conjugacy classes of elements in G. Classically we prove that χ is injective and $\mathrm{Im}(\chi)$ is a maximal lattice in the sense that:

$$R(G) \underset{\mathbb{Z}}{\otimes} \mathbb{C} \cong Cl(G).$$

This can be reformulated by identifying G with $\mathrm{Hom}(\mathbb{Z}, G)$. If G acts trivially on $\mathbb{C}$ and by conjugation on $\mathrm{Hom}(\mathbb{Z}, G)$, then:

$$Cl(G) \cong \mathrm{Map}_G(\mathrm{Hom}(\mathbb{Z}, G), \mathbb{C}).$$

Generalizing from $\mathbb{C}$ to an arbitrary field $\mathbb{F}$ and from $\mathbb{Z}$ to some larger discrete group Γ, we can define

$$C_{\Gamma,\mathbb{F}}(G) \text{ to be } \mathrm{Map}_G(\mathrm{Hom}(\Gamma, G), \mathbb{F}).$$

Note: An element of $C_{\Gamma,\mathbb{F}}(G)$ is a map $\mathrm{Hom}(\Gamma, G) \to \mathbb{F}$ that is constant on G-orbits.

LEMMA 3.1. *$C_{\Gamma,\mathbb{F}}()$ is a Mackey functor.*

We define induction as follows:

DEFINITIONS. If $H \subseteq G$, $\mathrm{ind}_H^G : C_{\Gamma,\mathbb{F}}(H) \to C_{\Gamma,\mathbb{F}}(G)$ is defined by $\mathrm{ind}_H^G(f)(\gamma) = 1/h \sum_{\substack{g \\ c_g\gamma \in C_{\Gamma,\mathbb{F}}(H)}} f(c_g \cdot \gamma)$, where $f : \mathrm{Hom}(\Gamma, H) \to \mathbb{F}$, $\gamma : \Gamma \to G$ and c_g denotes conjugation by g.

The properties of a Mackey functor now follow as they do for ordinary characters.

$\Gamma(G) =$ subgroups of G that occur as quotients of Γ. We make $\Gamma(G)$ into a category by defining the morphisms to be inclusions twisted by conjugation.

$I(G)$ equals the subcategory with morphisms given by inclusions only.

THEOREM 3.2. *The natural map $C_{\Gamma,\mathbb{F}} \xrightarrow{\simeq} \varprojlim_{H \in \Gamma(G)} C_{\Gamma,\mathbb{F}}(H)$ is an isomorphism.*

The same is obviously true for any category of subgroups containing $\Gamma(G)$.

Proof: $\varprojlim_{H \in I(G)} \mathrm{Hom}(\Gamma, H) \cong \mathrm{Hom}(\Gamma, G)$ from the definitions. Now let us take G-invariant maps into $\mathbb{F}$ noting that for any contravariant functor $\mathbf{F}$, $\varinjlim_{h \in \Gamma(G)} \mathbf{F}(H) = \left[\varinjlim_{h \in I(G)} \mathbf{F}(H) \right]^G$. (Here and elsewhere the superscript G denotes invariant elements.) ■

EXERCISE.

1. If Γ is abelian, $C_{\Gamma,\mathbb{F}}(G) = \lim_{\mathfrak{A}(G)} C_{\Gamma,\mathbb{F}}(A)$, where $\mathfrak{A}(G)$ denotes abelian subgroups of G.

2. If $\Gamma \cong \mathbb{Z}$, $C_{\mathbb{Z},\mathbb{F}}(G) = \lim_{\mathfrak{C}(G)} C_{\mathbb{Z},\mathbb{F}}(H)$, where $\mathfrak{C}(G)$ denotes cyclic subgroups of G. (Compare Artinian induction for ordinary representations.)

REMARK. If we work with $\mathbb{Z}$ or some finite extension of $\mathbb{Z}$ instead of with the complex numbers $\mathbb{C}$ as image $\mathbb{F}$, then we can deduce that:

$$R(G) \otimes \mathbb{Z}\left[\frac{1}{|G|}\right] \cong \lim_{\mathfrak{C}(G)} R(H) \otimes \mathbb{Z}\left[\frac{1}{|G|}\right].$$

That is, we must introduce denominators. We return to this point later.

3.3. Character Rings for $h^*(BG)$

Let $h^*(-)$ be a multiplicative cohomology theory of the kind discussed in Chap. 2. Then formal arguments imply that $G \rightsquigarrow h^*(BG)$ is a Mackey functor. (For K-theory and its variants this is almost immediate by the discussion in Sec. 3.1; see also Chap. 6.) We wish to study $h^*(BG)$ through characters:

$$\chi : h^*(BG) \to ?$$

just as we studied $R(G)$ and $Cl(G)$. To do this we reformulate the character map $\chi : R(G) \to Cl(G)$, noting that to give an element of G is equivalent to giving a map $g : \mathbb{Z}/N \to G$ for N equal to some multiple of the order of G. Qua rings $R(\mathbb{Z}/N) = \mathbb{Z}[x]/(x^N - 1)$; we can think of $\chi(g)$ as a composite map:

$$R(G) \underset{g^*}{\longrightarrow} R(\mathbb{Z}/N) \underset{\varphi}{\longrightarrow} \mathbb{C}$$

with φ given by $x \longmapsto e^{e\pi i/N}$.

Note: We can think of χ as a function of two variables, the module M and the group element g. The preceding composition allows M rather than g to vary. Thus we have a map:

$$\chi(g) : \text{“}R(G)\text{”} \to \mathbb{C},$$

where quotation marks refer to some generalized representation ring. Next replace the cyclic subgroup $\langle g \rangle$ by the abelian subgroup $\langle g_1, \ldots, g_n \rangle$ of G of rank n. Recall from Sec. 3.1 that we use $\mathbb{Z}_p$ to denote the p-adic integers:

$$\mathbb{Z}_p = \underset{n}{\overset{\lim}{\longleftarrow}} \{\mathbb{Z}/p \twoheadleftarrow \mathbb{Z}/p^2 \twoheadleftarrow \mathbb{Z}/p^3 \twoheadleftarrow \cdots \mathbb{Z}/p^n \twoheadleftarrow \cdots\}.$$

For some value of i, the map $\alpha : \mathbb{Z}_p^n \to G$ can be factored as:

$$\mathbb{Z}_p^n \longrightarrow (\mathbb{Z}/p^i)^n \underset{\alpha_i}{\longrightarrow} G.$$

Define $\chi(\alpha) : h^*(BG) \to \overline{\mathbb{Q}}_p$ to be the composition:

$$h^*(BG) \underset{(B\alpha_i)^*}{\longrightarrow} h^*B(\mathbb{Z}/p^i)^n \underset{\varphi_i}{\longrightarrow} \overline{\mathbb{Q}}_p,$$

subject to the provision that the appropriate compatible family of maps $\{\varphi_i\}$ exists. Here compatible means that for all $j > i$, the following diagram commutes

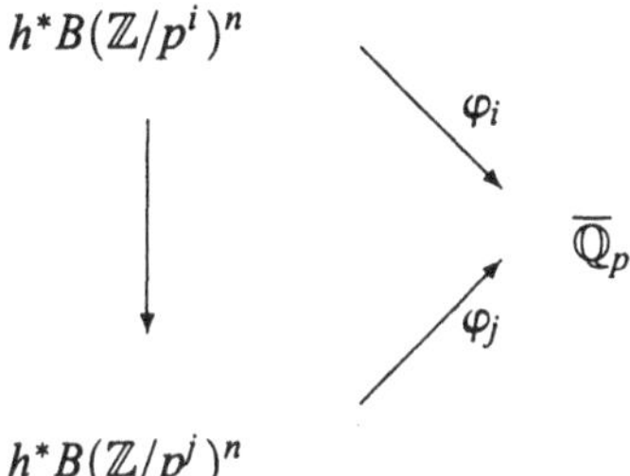

Topologically the conjugation map is such that $Bc_g \simeq 1$, so that $\chi(c_g\alpha) = \chi(\alpha)$ and χ is conjugation-invariant. Therefore once we find such a family of maps $\{\varphi_i\}$ for the theory $h^*(-)$, we have constructed a character:

$$\chi(\alpha) : h^*(BG) \longrightarrow \overline{\mathbb{Q}}_p$$

for each $\alpha : \mathbb{Z}_p^n \longrightarrow G$, which is invariant under the group action. Taking adjoints, this amounts to having a character:

$$\chi_G = \chi : h^*(BG) \to \mathrm{Map}_G(\mathrm{Hom}(\mathbb{Z}_p^n, G), \overline{\mathbb{Q}}_p)$$
$$\|$$
$$C_{\mathbb{Z}_p^n, \overline{\mathbb{Q}}_p}(G) \underset{DEF}{:=} Cl_{n,p}(G).$$

The p-adic numbers $\overline{\mathbb{Q}}_p$ are uneconomical, and in practice we make do with a much smaller subring, i.e., by adjoining elements and extending as various steps in the construction demand.

3.4. Lubin–Tate Construction

From now on let $h^*(-)$ be a complex-oriented cohomology theory. Using the same notation as in Chaps. 1 and 2, $h^*(\mathbb{C}P(\infty)) = h^*[[t]]$, $\deg t = 2$. The H-space structure $\mathbb{C}P(\infty) \times \mathbb{C}P(\infty) \to \mathbb{C}P(\infty)$ coming from the tensor product of line bundles gives rise to the formal group law $F_h = F$, and we write

$$[0](t) = 0$$
$$[m](t) = F(t, [m-1]t), \qquad m \geqslant 1.$$

The short exact sequence of groups $0 \to \mathbb{Z} \underset{p^i}{\to} \mathbb{Z} \to \mathbb{Z}/p^i \to 0$ induces a map of classifying spaces (fibration):

$$S^1 \to B(\mathbb{Z}/p^i) \to K(\mathbb{Z}, 2) = \mathbb{C}P(\infty).$$

The existence of Gysin maps in $h^*(-)$ leads to the following splitting:

$$0 \to h^*(\mathbb{C}P(\infty)) \to h^*(\mathbb{C}P(\infty)) \to H^*(B(\mathbb{Z}/p^i)) \to 0$$
$$1 \longmapsto [p^i](t).$$

Therefore $h^*(B(\mathbb{Z}/p^i)) \cong h^*[[t]]/[p^i](t)$, and more generally:

$$h^*(B(\mathbb{Z}/p^i)^n) \cong h^*[[t_1, \ldots, t_n]]/\langle [p^i](t_1), \ldots, [p^i](t_n)\rangle.$$

(The preceding is no more complicated than calculating the ordinary cohomology of a finite dimensional lens space.)

Write $\Lambda_i \subseteq \overline{\mathbb{Q}_p}$ for some fixed subgroup of $\overline{\mathbb{Q}_p}$ containing solutions of the equation $[p^i](t) = 0$, then assume that $h^*(-)$ has been p-localized. The theory of formal groups implies that:

$$[p](t) = u_n t^{p^n} + u_{n+1} t^{p^{n+1}} + \cdots, \qquad u_n \neq 0 \ (\mathrm{mod}\ p).$$

Definition. The leading nonzero coefficient u_n in the preceding equation determines the *height* n of the formal group law. We adopt the convention that if $[p](t) = 0$, then height $(F) = 0$.

Remark. The theory of elliptic curves E shows that the height of their formal groups is 1 or 2. Hence for a supersingular curve E (with $u_1 = 0$ necessarily), we must have $u_2 \neq 0$ (modulo p). This provides an alternative route to the regularity of $\{p, u_1, u_2\}$ Lemma 2.1 and its corollaries in Chap. 2. Note: Gross's result is needed to explain the role of the discriminant Δ.

Theorem 3.3 (Lubin–Tate). *Let $h^* \to \mathcal{O}$ be a local homomorphism into the ring of integers of a complete, algebraically closed, local field L. Suppose that the reduction of the formal group law F, modulo the maximal ideal of $\mathcal{O}$, has height n. Then $[p^i](x)$ has p^{ni} distinct zeros in L, which form a subgroup $\Lambda_i \cong (\mathbb{Z}/p^i)^n$. Furthermore as i increases, the isomorphisms are compatible, and:*

$$\Lambda_\infty = \lim_i \Lambda_i \cong (\mathbb{Z}_p)^n.$$

In the present context this result allows us to define a compatible family of maps $\{\varphi_i\} : h^*B(\mathbb{Z}/p^i) \to \overline{\mathbb{Q}}_p$ by mapping the power series variable t to some suitable root in Λ_i. Thus the structure of Λ_i provides our generalization of the ordinary character maps $x \mapsto e^{2\pi ri/n}$ and completes the construction of the characters:

$$\chi_G : h^*(BG) \to Cl_{n,p}(G).$$

Discussion of the proof of Theorem 3.3: Let E be a graded commutative ring, complete with respect to the topology defined by the powers of an ideal I. Suppose that $f(x) \in E[[x]]$ satisfies $f(x) \equiv ux^d \pmod{(I, x^{d+1})}$ with u a unit in E, then

1. (Euclidean algorithm) Given $g(x) \in E[[x]]$, there exist a power series $q(x)$ and a polynomial $r(x)$ of degree $\leqslant d-1$, such that $g(x) = q(x)f(x) + r(x)$.

   For a proof of this see [72, pp. 129–131]. (The idea behind this proof can also be found in [124, Vol. 2, pp. 139–141 and p. 261].)

   As consequences we have Results 2 and 3.

2. The quotient ring $E[[x]]/(f(x))$ is a free E-module on the basis $\{1, x, \ldots, x^{d-1}\}$.

3. There is a unique factorization $f(x) = u(x)f_1(x)$, where $f_1(x)$ is a monic polynomial of degree d (the Weierstrass degree of f).

If our formal group law F has height n, it now follows that $[p^i](x)$ has p^{in} zeros, *possibly with multiplicities*. Taking logarithms we have

$$\log_F([k](x)) = k \log_F(x).$$

Formally differentiating both sides:

$$\log_F'([k](x))[k]'(x) = k \log_F'(x). \tag{3.1}$$

For any logarithm of a formal group law, the formal derivative is a power series with initial term 1, hence invertible. Given (3.1) and the fact that $k \neq 0$, we see that $[k]'(x)$ has no zeros, implying that the zeros of $[k](x)$ are all distinct. The group of zeros Λ_i is thus as claimed. ∎

Note: The need to introduce Λ_i explains the use of $\overline{\mathbb{Q}}_p$ as image field for the character. But in practice and with a specific theory, such as elliptic cohomology, we can make do with a smaller range. If we are interested in groups of bounded order, we take this only large enough to contain the zeros of $[p^i](x)$ for some $i \leqslant N$.

The main theorem is best stated in terms of v_n-periodic cohomology theories — recall the discussion in Sec. 2.2. If n is the least integer such that $h^*(-)$ is v_n-periodic, then the formal group law has height n. Write $I_n \in h^*$ for the image of the ideal $\langle p, v_1, \ldots, v_{n-1}\rangle$.

THEOREM 3.4. *Let $h^*(-)$ be a multiplicative, complex-oriented, v_n-periodic cohomology theory, such that coefficients h^* are an integral domain (containing 1/2). Suppose there is a ring homomorphism $\varphi : h^* \to \overline{\mathbb{Q}}_p$ such that $\varphi(I_n)$ is contained in the maximal ideal of $\mathbb{Z}_p$, then:*

$$\chi_G : h^*(BG)^{\wedge}_{I_n} \underset{\widehat{h^*}}{\otimes} \overline{\mathbb{Q}}_p \to Cl_{n,p}(G)$$

is a natural isomorphism.

The notation $\hat{\cdot}$ refers to completion with respect to the I_n-adic topology. A little care is needed in replacing $\overline{\mathbb{Q}}_p$ by some smaller *subring*. For a general group G, we must be able to invert $|G|$. This can be avoided for some groups, such as those with periodic cohomology, for which we can show $h^*(BG)$ to be concentrated in even dimensions.

3.5. Reduction of the Theorem to the Cyclic Case

The main result follows from Propositions 3.1 and 3.2.

PROPOSITION 3.1. *χ_G is an isomorphism for all abelian groups G, and*

PROPOSITION 3.2. *χ_G is an isomorphism for all finite groups G if and only if χ_G is an isomorphism when G is abelian.*

In turn Proposition 3.2 follows from Proposition 3.3.

PROPOSITION 3.3. *Let $h^*(-)$ be a multiplicative, complex-oriented cohomology theory and G a finite group. Then the natural map:*

$$h^*(BG) \otimes \mathbb{Z}\left[\frac{1}{|G|}\right] \longrightarrow \lim_{A \in \mathfrak{A}(G)} h^*(BA) \otimes \left[\frac{1}{|G|}\right]$$

induced by restriction is an isomorphism.

Proof that Proposition 3.3 implies Proposition 3.2: Complete the isomorphism in Proposition 3.3 and tensor with $\overline{\mathbb{Q}}_p$, obtaining

$$h^*(BG)^\wedge \underset{\widehat{h}^*}{\otimes} \overline{\mathbb{Q}}_p \cong \lim_{A \in \mathfrak{A}(G)} h^*(BA)^\wedge \underset{\widehat{h}^*}{\otimes} \overline{\mathbb{Q}}_p.$$

Correspondingly we already proved (Theorem 3.2) that

$$Cl_{n,p}(G) \cong \lim_{A \in \mathfrak{A}(G)} Cl_{n,p}(A),$$

and the construction of χ_G is natural with respect to restriction. The category $\mathfrak{A}(G)$ suffices because we suppose that Γ is free-abelian and we are concerned with quotients of Γ in G. ■

Proof of Proposition 3.1: We know that complex orientation and the formal group law determine the structure of $\widehat{h}^*(BA)$ as a quotient of $\widehat{h}^*[[t_1, \ldots, t_k]]$. From the earlier discussion (Note: $\widehat{h}_*$ is now complete!) it follows that $\widehat{h}^*(B\mathbb{Z}/p^i)$ is free of rank p^{ni} on elements $\{1, t, \ldots, t^{p^{ni}-1}\}$. Freeness means that $\widehat{h}^*$ takes a product of cyclic groups into a tensor product; hence to prove Proposition 3.1, it suffices to assume that $A \cong \mathbb{Z}/p^i$. The space of class functions:

$$Cl_{p,n}(\mathbb{Z}/p^i) = \mathrm{Map}_{\mathbb{Z}/p^i}(\mathrm{Hom}(\mathbb{Z}_p^n, \mathbb{Z}/_{p^i}), \overline{\mathbb{Q}}_p)$$

is likewise free with the same rank p^{ni}. Hence we must show that the $p^{ni} \times p^{ni}$ matrix representing $\chi_{\mathbb{Z}/p^i}$ is nonsingular. Let $\{e_1, \ldots, e_n\}$ be a basis for $\Lambda_i \cong (\mathbb{Z}/p^i)^n$, t the multiplicative generator for $h^*(B\mathbb{Z}/p^i)$ and $(a_1, \ldots, a_n)$ some n-tuple of elements from $\mathbb{Z}/p^i$ describing an element of $Cl_{p,n}(\mathbb{Z}/p^i)$, then

$$\chi(a_1, \ldots, a_n)(t) = [a_1](e_1) \underset{F}{+} \cdots \underset{F}{+} [a_n](e_n) \in \Lambda_i \subseteq \overline{\mathbb{Q}}_p.$$

Therefore if $z_1, \ldots, z_{p^{in}}$ are the distinct zeros in Λ_i, the matrix representing $\chi_{\mathbb{Z}/p^i}$ is the Vandermonde matrix (z_j^l). Since all the z_j are distinct this matrix has nonzero determinant. ■

3.6. Proof of Proposition 3.3

The full details of this proof require a knowledge of equivariant homotopy theory. However Proposition 3.3 itself has the flavor of Artin's induction theorem; hence it should at least appear plausible to algebraists. A good background reference is [77].

Definition. The *Burnside Ring* $A(G)$ of a finite group is the Grothendieck group associated with finite G-sets. These can be added using disjoint union, and multiplied by taking the Cartesian product with diagonal action. Equivalently it is the Grothendieck group on isomorphism classes of transitive G-sets G/H. The nontriviality of the product is essentially the double-coset formula.

There is a natural isomorphism $A(G) \cong \{S^0, S^0\}_G$, G-equivariant stable maps from S^0 to S^0, and the definition in terms of homotopy theory leads to an $A(G)$-action on any equivariant cohomology theory. We can define a character map for $A(G)$: Given $H \in G$, let

$\chi_H : A(G) \to \mathbb{Z}$ be given by $\chi_H(X) = |X^H|$ = number of points in X fixed by H.

Lemma 3.2. *The map:*

$$\prod_{\substack{\text{conj. classes}\\ \text{of subgps. } H \subseteq G}} \chi_H : A(G) \to \prod_{\substack{\text{conj. classes}\\ \text{of subgps. } H \subseteq G}} \mathbb{Z}$$

is injective. Inversion of $|G|$ *makes it surjective.*

(This explains the appearance of $1/|G|$ in Proposition 3.3.)
It follows that we can write $1 \in A(G) \otimes \mathbb{Z}[1/|G|]$ as $1 = \Sigma_{(H)} e_H$, with:

$$\chi_K(e_H) = \begin{cases} 1 & \text{if } K \text{ is conjugate to } H \\ 0 & \text{otherwise} \end{cases}$$

Therefore $e_H \in A(G)$ is idempotent, and given any additive, contravariant functor h : G-complexes $\rightsquigarrow \mathbb{Z}[1/|G|]$-modules, the $A(G)$-action on $h(-)$ and the relation $1 = \Sigma_{(H)} e_H$ give

$$h(X) = \prod_{(H)} e_H h(X).$$

Indeed we have Proposition 3.4.

Proposition 3.4 (J. F. Adams).

$$h^*(EG \underset{G}{\times} X) \otimes \mathbb{Z}\left[\frac{1}{|G|}\right] \xrightarrow{\sim} \prod_{(H)} e_H h^*(BH \times X^H)^{W_G(H)} \otimes \mathbb{Z}\left[\frac{1}{|G|}\right]$$

is an isomorphism, where on the right-hand side X *is regarded as an* H*-space. Invariance is with respect to the Weyl group* $W_G(H) = N_G(H)/H$*, which acts on the right of* $G/H \times X^H$ *by* $(gH, x)(nH) = (gnH, n^{-1}x)$.

Again with ordinary cohomology and representation theory in mind, this is the kind of result we expect.

Recall the following key lemma from the proof of the splitting principle used in Chap. 1 for complex vector bundles.

LEMMA 3.3. *Let ξ be a $\mathbb{C}^n$-bundle over Y and $\pi : P(\xi) \to Y$ the associated $\mathbb{C}P^{n-1}$-bundle. Then for any complex-oriented cohomology theory $h^*(-)$, $h^*(P(\xi))$ is a free module of rank n over $h^*(Y)$ with the module action defined using π^*. In particular π^* is injective.*

Proof: Given the structure of $h^*(\mathbb{C}P^{n-1})$, this is clear by the Künneth formula for trivial bundles. Now use a Mayer–Vietoris argument. ■

Embed G is some unitary group $U(m)$, then let F (for flag) be the left G-space $U(m)/T^m$. The map $F \to *$ is the composite of $(m-1)$ projective complex-bundle projections and hence the same holds for:

$$\pi : EG \underset{G}{\times} (X \times F) \longrightarrow EG \underset{G}{\times} X.$$

COROLLARY 3.1. *For any complex-oriented cohomology theory:*

$$\pi^* h^*(EG \underset{G}{\times} X) \longrightarrow h^*(EG \underset{G}{\times} (X \times F))$$

is injective.

This enables us to prove Proposition 3.3 by first solving the problem in $EG \underset{G}{\times} (X \times F)$.

We have the commutative diagram:

$$\begin{array}{ccc} h^*(EG \underset{G}{\times} X) \otimes \mathbb{Z}\left[\frac{1}{|G|}\right] & \xrightarrow[-]{\sim} & \prod_{(H)} e_H h^*(BH \times X^H)^W \otimes \mathbb{Z}\left[\frac{1}{|G|}\right] \\ \downarrow \pi_G^* & & \downarrow \prod \pi_H^* \\ h^*(EG \underset{G}{\times} (X \times F)) \otimes \mathbb{Z}\left[\frac{1}{|G|}\right] & \xrightarrow[-]{\sim} & \prod_{(H)} e_H h^*(BH \times X^H \times F^H)^W \otimes \mathbb{Z}\left[\frac{1}{|G|}\right] \end{array}$$

where the vertical maps are injective. The quotient space $U(m)/T^m$ has a fixed point *if and only if* H is conjugate to a subgroup of T^m, that is *if and only if* H is abelian. Hence $h^*(BH \times X^H \times F^H)$ is trivial unless H is abelian. The injectivity of π^* implies that $e_H h^*(BH \times X^H)^W$ is also trivial unless H is abelian. With X equal to a point, we conclude that:

$$h^*(BG) \otimes \mathbb{Z}\left[\frac{1}{|G|}\right] \xrightarrow[-]{\sim} \prod_{\mathfrak{A}(G)} e_A h^*(BA)^W \otimes \mathbb{Z}\left[\frac{1}{|G|}\right].$$

If A is abelian, write $\mathbf{F}(A) = h^*(BA) \otimes \mathbb{Z}[1/|G|]$. Then if A' is a subgroup of A we have the commutative diagram:

$$\begin{array}{ccc} \mathbf{F}(A) & \xrightarrow[=]{\sim} & \prod_{A'' \subseteq A} e_{A''} \mathbf{F}(A'') \\ \big\downarrow \text{res} & & \big\downarrow \\ \mathbf{F}(A') & \xrightarrow[=]{\sim} & \prod_{A'' \subseteq A'} e_{A''} \mathbf{F}(A'') \end{array},$$

where the right vertical map is a projection onto some of the factors. It follows that there is a natural isomorphism $\prod_{A \subseteq G} e_A \mathbf{F}(A) \cong \lim_{I(G)} \mathbf{F}(A)$. Taking G-invariant elements, we have

$$\prod_{(A)} e_A \mathbf{F}(A)^W \cong \left[\prod_{A \subseteq G} e_A \mathbf{F}(A) \right]^G \cong \left[\lim_{I(G)} \mathbf{F}(A) \right]^G = \lim_{\mathfrak{A}(G)} \mathbf{F}(A),$$

concluding the proof of Proposition 3.3.

3.7. Notes

Chapter 3 is based largely on notes circulated by J. H. Hunton as part of a seminar held at Cambridge in spring 1992. These notes are based on various preprints circulated as [56]. We are grateful to Hunton for allowing us to use these here. The theory they develop is essentially local; in terms of the original Atiyah theorem, the main result should be seen as the analog of the isomorphism:

$$(K\mathbb{Z}_p)^*(BG) \cong R(G) \otimes \mathbb{Z}_p.$$

However J. P. Greenlees shows that a global version can be established, providing yet more motivation for a geometric definition of $Ell^*(X)$ over the integers.

The following points clarify the position that Proposition 3.4 is a generalization of Artin's induction theorem for rational representations (see the proof given in [108]):

1. The natural equivariant map $\pi : G/H^+_\wedge X^H \longrightarrow X$ induces an isomorphism $e_H \cdot h(X) \cong e_H \cdot h(G/H^+_\wedge X^H)^W$, where as usual the affix $+$ indicates adjunction of a disjoint base point. This depends on the bijectivity of a fixed-point map, which reduces to the decomposition of the unit element given in Lemma 3.2 if $X =$ point. It is proved by induction using a five-lemma argument based on equivariant cells.

2. Quillen's descent to the flag manifold argument allows us to restrict to abelian subgroups A (compare Corollary 3.1).

3. Note: Taking limits over the category $\mathfrak{A}(G)$ is equivalent to taking limits over $I(G)$ and restricting attention to invariant elements.

Hopkins, Kuhn, and Ravenel also prove a formula for evaluating the induction map $i_* : h^*BH \to h^*BG$ for the inclusion $i : H \hookrightarrow G$. After completion, taking tensor products, etc., we obtain

$$\chi_G(g_1, \ldots, g_n)(i_* x) = \sum_{\substack{\text{invariant} \\ Hg \in G/H}} \chi_H(g g_1 g^{-1}, \ldots, g g_n g^{-1})(x)$$

In Chap. 7 we reinterpret Theorem 3.4 in terms of equivariant elliptic cohomology. Like the argument given in Chap. 3, this hinges on the fact that if we have the correct coefficients, we can reduce to the case of cyclic groups, for which the required isomorphism comes with the complex orientation.

4

Mathieu Groups

Most concepts considered so far can be illustrated by the sporadic Mathieu group M_{24}. Eventually we hope to replace M_{24} by the Monster group $\mathbb{M} = F_1$, but in the meantime M_{24} has the advantage of being large enough to reflect some of the flavor of the general case yet small enough to be tractable. Chapter 4, a middle ground between comparatively well-known material and the more specialized treatment of classifying spaces, also introduces a family of groups that play an important if still rather mysterious role later on.

4.1. Construction of Mathieu Groups

The five Mathieu groups were originally constructed as multiply transitive permutation groups in S_{12} and S_{24}. In terms of specific permutations, $M_{12} \subseteq S_{12}$ is generated by $(1 \cdots 11), ((5,6,4,10),(11,8,3,7))$, and $((1,12)$, $(2,11)$, $(3,6)$, $(4,8)$, $(5,9)$, $(7,10))$ and

$$M_{24} \subseteq S_{24}$$

by the permutations $(1 \cdots 23), ((3,17,10,7,9),(5,4,13,14,19),(11,12,23,8,18),(21,16,15,20,22))$, and $((1,24)$, $(2,23)$, $(3,12),(4,16)$, $(5,18)$, $(6,10)$, $(7,20)$, $(8,14)$, $(9,21)$, $(11,17)$, $(13,22)$, $(19,15))$.

The remaining three groups $M_{11} \subseteq M_{12}$ and $M_{22}, M_{23} \subseteq M_{24}$ occur as stabilizer subgroups; we describe these permutations only because of the cycle decomposition for various conjugacy classes used later. A more enlightening description is stated in terms of automorphisms of Steiner systems $S(k,m,n)$ consisting of m-ads (blocks) in a set of n elements. These are such that each k-ad lies in one and only one m-ad. We have a chain of inclusions:

$$S(2,5,21) \subset S(3,6,22) \subset S(4,7,23) \subset S(5,8,24),$$

with automorphism groups $L_3(4)$, M_{22}, M_{23}, and M_{24}, respectively. Here $L_3(4)$ denotes the automorphism group of the smallest Steiner system just given, isomorphic to the projective plane $\mathbb{F}_4P^2$. (A line is uniquely determined by two of its points.) An inductive argument group by group shows the simplicity of the Mathieu family. Similar inclusions:

$$S(3,4,10) \subset S(4,5,11) \subset S(5,6,12)$$

allow us to define M_{11} and M_{12}. The automorphism group M_{10} of $S(3,4,10)$ contains a unique minimal normal subgroup isomorphic to A_6 ($\cong PSp_4(2)$). The group M_{12} embeds in M_{24} as the stabilizer of a 12-ad; the 6-ads of the smaller system arise from the 8-ads meeting this 12-ad in 6 points. Furthermore the nontrivial outer automorphism of M_{12} interchanges the chosen 12-ad with its complement in the set of 24 elements. All this is described in [35]; see also [12, Chap. 6].

However a few words of explanation are necessary. By deleting a point from the set Ω of order n, it is clear that we obtain a Steiner system $S(k-1,m-1,n-1)$, and extension reverses this residual construction. In passing from $S(k,m,n,)$ to $S(k+1,m+1,n+1)$ the trick is to choose the right definition of the new blocks, each of which contain $m+1$ elements from the extended set $\Omega' = \{z\} \cup \Omega$. We start with the notion of *independence* in Ω; for $S(k,m,n)$ this requires that no $(t+1)$-ad be contained in a block B of Ω. Blocks of the extension are then of two kinds:

1. $\{z\} \cup B$, where B is a block in Ω.
2. Independent $(k+1)$-ads B' of Ω such that each independent $(t+1)$-ad is contained in a unique B'.

Constructing the two Mathieu chains now requires us to analyze the independent subsets of $\mathbb{F}_4P^2$. The group $PSL_3(\mathbb{F}_4)$ acts on the set of independent hexads, dividing it into three orbits, one of which provides the extension subset Type 2 above. The group M_{22} is then defined to be the group of alternating automorphisms of the extended Steiner system with 22 elements. The 2-transitive property of L_3 implies the 3-transitive property for M_{22}. Repeating the construction gives M_{23} and M_{24}, at which point the process stops, since the 24-element set has no independent heptads (see [12, Sec. 18] for more details).

By exploiting k-transitivity ($k = 1,3,4,5$), and at each stage defining G_p to be the subgroup of the automorphism group fixing a point p, we can write

$$G = KG_p(1 \neq K \lhd G).$$

The simplicity of $L_3(4)$ now easily implies that of M_{22}, M_{23}, and M_{24}.

There is yet another description in terms of automorphisms of finite projective planes, thus:

$$M_{12} = \langle L_2(11), f\rangle, \qquad f(x) = 4x^2 - 3x^7,$$
$$M_{24} = \langle L_2(23), g\rangle, \qquad g(x) = -3x^{15} + 4x^4.$$

Instead of g we can use the automorphism $h(x) = 9^{\varepsilon}x^3$, where $\varepsilon = \pm 1$ depending on whether x is a square or a nonsquare in $\mathbb{F}_{23}$. For our purposes representations

$$C : M_{12} \to PGL_5(3)$$
$$T : M_{24} \to GL_{11}(2)$$

are particularly important (see [116, 117]). In the case of M_{24}, if we lift to characteristic zero, we obtain the explicit virtual character:

1^{24}	$1^6 3^6$	$1^4 5^4$	$1^3 7^3$	$1^3 7^3$	$1^2 11^2$	
11	2	1	$\frac{1-i\sqrt{7}}{2}$	$\frac{1+i\sqrt{7}}{2}$	0	
$1\ 15\ 5\ 3$	$1\ 15\ 5\ 3$	$1\ 23$	$1\ 23$	3^8	$21\ 3$	$21\ 3$
$\frac{-1+i\sqrt{15}}{2}$	$\frac{-1-i\sqrt{15}}{2}$	$\frac{-1+i\sqrt{23}}{2}$	$\frac{-1-i\sqrt{23}}{2}$	-1	$-1+i\sqrt{7}$	$-1-i\sqrt{7}.$

For reference we note that the remaining (2-singular) conjugacy classes are given by:

$1^8 2^8 \quad 1^4 4^4 2^2 \quad 1^2 8^2 4\ 2 \quad 1^2 6^2 3^2 2^2 \quad 1\ 14\ 7\ 2$ (twice)

$12^2 \quad 6^4 \quad 4^6 \quad 2^{12} \quad 10^2 2^2 \quad 4^4 2^4 \quad 12\ 6\ 4\ 2.$

Here permutations are used to describe the classes; i.e., $(1)^{j_1}\ (2)^{j_2} \cdots (r)^{j_r}$ denotes the product of j_1 cycles of length 1, j_2 cycles of length 2, $\cdots$ with $j_1 + 2j_2 \cdots + rj_r = 24$. For example generating cycles on page 49 above have types $1\ 23$, $1^4 5^4$, and 2^{12}.

4.2. Conjugacy Classes and Modular Forms according to G. Mason

In Chap. 6 we give an algebraic description of $Ell^*(BM_{24})$, that shows the importance of the Todd representation T, and we also show how certain infinite dimensional vector bundles are naturally associated with functions in the class groups $Cl_{2,p}M_{24}$ as p runs through the primes dividing the order. In this section we use ordinary characters to describe the simplest of these bundles, which seem to be of interest from several points of view.

Recall from Chap. 2 that the ring of coefficients for one version of elliptic cohomology may be identified with $\mathbb{Z}[1/2][\delta,\varepsilon,\varepsilon^{-1}]$, where δ and ε are modular forms of weights 2 and 4, respectively, and the invariance subgroup is $\Gamma_0(2)$ contained in $PSL_2(\mathbb{Z})$. Formally and allowing other invariance subgroups we have the following definitions.

DEFINITION. Let $k \in \mathbb{Z}$ and $\Gamma \subseteq PSL_2(\mathbb{Z})$ be of finite index. A function $f : H_1 \to \mathbb{C}$ from the upper-half plane to the complex numbers is called a modular form of weight k with invariance subgroup Γ if:

1. f is holomorphic on H_1.
2. $f(z) = (cz+d)^{-k} f(Az)$ for all $A = \left(\begin{smallmatrix} a b \\ c d \end{smallmatrix}\right) \in \Gamma$ and $z \in H_1$.
3. $(cz+d)^{-k} f(Az)$ has a Fourier expansion in nonnegative powers of $e^{2\pi i z/N}$ for all $A \in PSL_2(\mathbb{Z})$ and some suitable natural number N.

In what follows we allow Γ to be a congruence subgroup $\Gamma_0(N)$, where $c \equiv 0 \pmod{N}$, and we also allow twisting of Relation 3 by means of a Dirichlet character $\varepsilon \pmod{N}$. We do not emphasize this last point, since in the cases of most interest to us ε is trivial. Somewhat informally we can think of the cusps of the complex quotient surface $\Gamma \backslash H_1$ as additional points needed in compactification.

DEFINITION. A modular form that vanishes at each cusp is called a cusp form.

These definitions are considered at greater length in [53, 95]. Denote the space of cusp forms of level N, weight k, and character ε by $S(N,k,\varepsilon)$. This space admits an action by the ring of Hecke operator $T(n)$ via the formula $(\varepsilon = 1)$

$$T(n)f(z) = n^{k-1} \sum_{\substack{a \geqslant 1, ad=n \\ 0 \leqslant b < d}} d^{-k} f\left(\frac{az+b}{d}\right).$$

The Hecke operators are analogous to cohomology operations, and more revealingly than the preceding formula, these satisfy the relations:

$$T(m)T(n) \quad \text{if } (m,n) = 1$$
$$T(p)T(p^n) = T(p^{n+1}) + p^{k-1}T(p^{n-1}) \qquad \text{for } p \text{ prime and } n \geqslant 1.$$

Thus they are multiplicative, and we ask for conditions under which a form is $T(n)$-invariant for all n.

Denote the compactified fundamental region for the subgroup $\Gamma_0(N)$ in the upper-half plane H_1 by $X_0(N)$. For a form $f \neq 0$ in $S(N,k,\varepsilon)$, denote the order of zero of f at a point z of $X_0(N)$ by $\operatorname{ord}(z,f)$. Write μ equal to the index of $\Gamma_0(N)$

in $\Gamma_0(1)$. A counting argument [95, p. 26 and Appendix C] shows that:

$$\mu = N\prod_{p|N}\left(1+\frac{1}{p}\right).$$

Integrating f'/f round the boundary of a suitable domain [95, Theorem 4.1.4, pp. 98–101] shows that:

$$\sum_{z\in X_0(N)} \operatorname{ord}(z,f) = \frac{\mu k}{12}.$$

Finally the number of cusps of the compactified quotient surface can also be counted, giving

$$\kappa(N) = \sum_{d|N}\varphi\left(\left(d,\frac{N}{d}\right)\right),$$

where φ is Euler's phi function [95, Sec. 2.4].

Restricting attention to forms of even weight with no twisting character ($\varepsilon = 1$), [95, Corollary 4.1.2] implies that ord (ξ,f) is an integer for each cusp z in $X_0(N)$. Imposing the condition $\kappa(N) = \mu k/12$ then forces $\operatorname{ord}(z,f) = 1$ at each cusp z for each nonzero form $f \in S(N,k,1)$. There can be no other zeros and necessarily $\dim S(N,k,1) = 1$, showing that a generating form is $T(n)$-invariant for all n. The preceding relation admits only finitely many solutions (N = level, k= weight), which are listed in Table 4.1.

The corresponding basic forms can be written out in terms of the indicated permutation from S_{24}. Indeed if:

$$\eta(z) = q^{1/24}\prod_{r\geqslant 1}(1-q^r), \qquad q = e^{2\pi i z},$$

the conjugacy class of g is described by the permutation $(1)^{j_1}(2)^{j_2}\cdots$, whose basic form is

$$\eta_g(z) = \eta(z)^{j_1}\eta(2z)^{j_2}\cdots$$

Here we define a conjugacy class to be even if the total number of cycles is divisible by 4. We use the same rule to define $\eta_g(z)$ for the remaining odd classes; thus for example the class $1^3 7^3$ is associated with the form $\eta(z)^3\eta(7z)^3$, it has weight 3 and level 7, modulo twisting by a nontrivial ε_g.

If we write $\eta_g(z) = \sum_{n=1}^{\infty}\theta_n(g)q^n\,(q = e^{2\pi i z})$, using the Fourier expansion for $\eta(z)$, and allow g to run through all conjugacy classes, we obtain an infinite dimensional (virtual) representation module $\Theta = \{\theta_n : n \geqslant 1\}$ for M_{24} (see [81]).

Table 4.1. Even Conjugacy Classes in M_{24}

N	k		N	k	
1	12	1^{24}	11	2	$1^2 11^2$
2	8	$1^8\,2^8$	12	2	1 2 7 14
3	6	$1^6\,3^6$	15	2	1 3 5 15
4	6	2^{12}	20	2	$2^2 10^2$
5	4	$1^4\,5^4$	24	2	2 4 6 12
6	4	$1^2\,2^2 3^2 6^2$	36	2	6^4
8	4	$2^4\,4^4$	27^a	2	$3^2 9^2$
9	4	3^8	32^a	2	$4^2 8^2$

[a]These entries do not correspond to even conjugacy classes in M_{24}.

We present evidence for this in Secs. 4.3 and 8.3. However the real bonus for choosing the $\eta_g(z)$ to be eigenforms is that we can expand the formal L-series $L(s) = \sum_{n=1}^{\infty} \theta_n n^{-s}$, which converges for $\mathrm{Re}(s) > k$, as an infinite product:

$$\prod_p \left[1 - \frac{\theta_p(g)}{p^s} + \frac{\varepsilon_g(p)p^{k(g)-1}}{p^{2s}}\right]^{-1}.$$

Because this expansion brings us back to elliptic genera, it is worth isolating in a separate section.

4.3. Eigenforms for Hecke Operators

Let $f(z) = \sum_{n=0}^{\infty} a(n)q^n$ be an eigenform of even weight k, which is not identically zero, and write $T(n)f = \lambda(n)f$ for all $n \geqslant 1$.

LEMMA 4.1.

1. $a(1) \neq 0$.
2. *If* $a(1) = 1$, *then* $a(n) = \lambda(n)$ *for* $n > 1$.

Sketch of proof: Direct calculation, starting with the defining equation for $T(n)f$, shows that $a(n)$ equals the coefficient of q in $T(n)f$. The eigenvalue condition also shows that it equals $\lambda(n)a(1)$. However if $c(1) = 0$, all coefficients of f vanish, contradicting nontriviality. Now normalize with $a(1) = 1$.

Note: Lemma 4.1 implies that two eigenforms with the same eigenvalues and $a(1) = 1$, actually coincide. ■

Table 4.2. Conjugacy Classes and Elliptic Curves

$1^2 11^2$	y^2		$+y$	$=x^3$	$-x^2$	$-10x$	-20	$11B^b$
$1\,2\,7\,14$	y^2	$+xy$	$+y$	$=x^3$		$+4x$	-6	$14C$
$1\,3\,5\,15$	y^2	$+xy$	$+y$	$=x^3$	$+x^2$	$-10x$	-10	$15C$
$2^2 10^2$	y^2			$=x^3$	$+x^2$	$+4x$	$+4$	$20B$
$2\,4\,6\,12$	y^2			$=x^3$	$-x^2$	$-4x$	$+4$	$24B$
$3^2 9^{2a}$	y^2		$+y$	$=x^3$			-7	$27B$
$4^2 8^{2a}$	y^2			$=x^3$		$+4x$		$32B$
6^4	y^2			$=x^3$			$+1$	$36A$

[a]These entries do not correspond to even conjugacy classes in M_{24}.

[b]The entry in Column 3 refer to tables in 1–3 [21]

LEMMA 4.2. *Under the same conditions as before, if* $(m,n)=1$, $a(m)a(n)=a(mn)$, *and if* p *is prime* $a(p)a(p^n)=a(p^{n+1})+p^{k-1}a(p^{n-1})$.

Proof: Obvious. ∎

LEMMA 4.3. *The L-series* $L_f(s)=\prod_p[1-a(p)p^{-s}+p^{k-1-2s}]$.

Sketch of proof: Reduce to a single factor by using the L-series relation $L(s)=\prod_p\left[\sum_{n=0}^{\infty}a(p^n)p^{-ns}\right]$. Now compare coefficients on both sides of:

$$\left[\sum_n a(p^n)p^{-ns}\right][1-a(p)p^{-s}+p^{k-1-2s}]=1$$

then use Lemma 4.2. ∎

Returning to the infinite dimensional representation Θ, specialize to those classes having $k(g)=2$.

The L-series are associated with elliptic curves over $\mathbb{Q}$ (see [52]). They can be listed (see [79]), and all turn out to be strong Weil curves with a finite Mordell–Weil group. Table 4.2 continues Table 4.1.

Note: For odd primes other than 23 dividing $|M_{24}|$ these classes are enough to detect Sylow subgroups; replacing $\eta_{(1,23)}$ by its square, we can bring even this prime, rather artificially, into the picture.

In the next subsection, we write the equation of each of these curves in the standard form:

$$Y^2=4X^3-g_2X-g_3=4(X-e_1)(X-e_2)(X-e_3).$$

With such a curve we can associate a lattice L, which contains a sub-lattice $\widetilde{L}$ of index 2 with respect to which the function $f(z):=1/\sqrt{\mathfrak{P}(z)-e_1}$ is elliptic.

Furthermore $f'(z) = [1 - 2\delta f(z)^2 + \varepsilon f(z)^4]$, where $\delta = -3e_1/2$ and $\varepsilon = (e_1 - e_2)(e_1 - e_3)$.

Thus each of the curves just listed gives rise to an elliptic genus. Stated in a way more in line with Mason's original treatment, the composition rule for each curve defines a formal group law, which is equivalent to the formal group law associated with the genus φ_g by the method in Sec. 1.2. To summarize the argument so far, cycle types of conjugacy classes in M_{24} are naturally associated with modular forms η_g, with the link provided by T_n-invariance. As g varies the forms $\{\eta_g\}$ define the character of a graded infinite dimensional representation. In about half the cases, coefficients of the modular form determine an elliptic curve over $\mathbb{Q}$ whose multiplication is equivalent to a genus. Two notes of warning are necessary:

1. The *level* of the modular form η_g can be much greater than 2.
2. Although there is a genus associated with forms of *weight* greater than 2, this need no longer be elliptic.

On the positive side, as we see in a later chapter, the theory extends to conjugacy classes of permuting pairs, and the resulting family of representations $\{\theta_{n,h}(g)\}$ fits well into the conjectural geometric framework for $Ell^*(BG)$. The definition must however be extended to cover forms of level N; at least formally it is clear that this can be done.

4.4. Eight Elliptic Genera of Mathieu Type

For the polynomials appearing below we have the following factorizations, given in increasing order of difficulty:

6^4

$$y^2 = x^3 + 1$$
$$(2y)^2 = 4(x+1)(x+\omega)(x+\omega^2) \quad \omega = e^{2\pi i/3}$$

$4^2 8^2$ [a]

$$y^2 = x^3 + 4x$$
$$(2y)^2 = 4(x^3 + 4x)$$
$$= 4(x(x+2i)(x-2i))$$

$3^2 9^2$ [b]

$$y^2 + y = x^3 - 7$$
$$(2y+1)^2 = 4x^3 - 27$$

[a] See Table 4.1.
[b] See Table 4.2.

$$= 4(x-\lambda)(x-\lambda\omega)(x-\lambda\omega^2) \quad \omega = e^{2\pi i/3},\ \lambda = 3/\sqrt[3]{4}$$

2 4 6 12
$$y^2 = x^3 - x^2 - 4x + 4$$
$$(2y)^2 = 4(x-1)(x-2)(x+2)$$

$2^2 10^2$
$$y^2 = x^3 + x^2 + 4x + 4$$
$$(2y)^2 = 4(x+1)(x+2i)(x-2i)$$

1 3 5 15
$$y^2 + xy + y = x^3 + x^2 - 10x - 10$$
$$(2y+x+1)^2 = 4[(x+1)(x^2-10)] + (x+1)^2$$
$$(4y+2x+2)^2 = 4[(x+1)(x-3)(4x+13)]$$

1 2 7 14
$$y^2 + xy + y = x^3 + 4x - 6$$
$$(2y+x+1)^2 = 4x^3 + x^2 + 18x - 23$$
$$= (x-1)(4x^2+5x+23)$$

The quadratic factor splits as:

$$\left[x + \frac{5 - i(7)^{3/2}}{8}\right]\left[x + \frac{5 + i(7)^{3/2}}{8}\right].$$

$1^2 11^2$
$$y^2 + y = x^3 - x^2 - 10x - 20$$
$$(2y+1)^2 = 4(x^3 - x^2 - 10x - 20) + 1$$

Factorizing the right-hand side using Maple gives

$$(x-\alpha)(x-\beta)(x-\overline{\beta}),$$

where $\alpha = 4.346308158$ and $\beta = -1.673154079\cdots + 1.320848922\cdots i$; that is the cubic equation has no straightforward solution, as in the first seven cases.

Passing from a Weierstrass to a Jacobi equation allows us to write down an elliptic genus, as described in Sec. 2.1.

EXAMPLE. Consider the class $2^2 10^2$. Completing the cube in order to express $4(x^3+x^2+4x+4)$ in Weierstrass form, i.e., changing coordinates to ensure that $e_1+e_2+e_3=0$, replaces the triple $(-1, 2i, -2i)$ by $(-2/3, 1/3+2i, (1/3)-2i)$. Therefore $\delta = -3e_1/2 = 1$ and:

$$\varepsilon = \left(\frac{1}{3} - 2i\right)\left(\frac{1}{3} + 2i\right) = \frac{37}{9}.$$

An obvious question is do elliptic genera constructed in this way have a geometric significance analogous to the degenerate cases of L and $\hat{A}$?

4.5. Thompson Series and Ramanujan Numbers

We return to the graded, infinite-dimensional representation $\Theta = \{\theta_n : n \geqslant 1\}$ of M_{24}, whose character at the class g is given by $\eta_g(z)$. The formal equality:

$$\sum_{n=1}^{\infty} \theta_n n^{-s} = \prod_p \left[1 - \frac{\theta_p}{p^s} + \frac{\varepsilon(p)p^{k-1}}{p^{2s}}\right]^{-1}$$

implies that the sequence $\{\theta_n : n \geqslant 1\}$ is multiplicative, i.e.:

1. $\theta_n \theta_m = \theta_{nm}$ if m and n are coprime.
2. $\theta_{p^{n+1}} = -\varepsilon(p)p^{k-1}\theta_{p^{n-1}} + \theta_p \theta_{p^n}$.

We recall also that the Dirichlet character ε is trivial provided k is even. In particular evaluating at the identity gives the relation between dimensions:

$$\theta_{p^{n+1}}(1) = -p^{11}\theta_{p^{n-1}}(1) + \theta_p(1)\theta_{p^n}(1).$$

Since $\eta_1(z)$ is the discriminant $\Delta(z)$, the sequence of coefficients $\{\theta_n(1), n \geqslant 1\}$ coincides with the Ramanujan sequence $\{\tau(n), n \geqslant 1\}$ with the additional information that $\tau(n)$ is the degree of a virtual representation of M_{24}. In principle at least we can say much more. In Chap. 8 we obtain a formula for summands of the Thompson series Θ in terms of the natural representation space V on which M_{24} acts by permutations. If $\wedge^r(V)$ denotes the rth exterior power of V and we abbreviate $\wedge^r(V)$ as λ^r, then:

$$\begin{aligned}
\Theta &= q\prod_{k=1}^{\infty}\sum_{r=0}^{\infty}(-1)^r \wedge^r(V) q^{kr} \\
&= q\prod_{k=1}^{\infty}(1 - \lambda^1 q^k + \lambda^2 q^{2k} - \lambda^3 q^{3k} + \cdots) \\
&= q(1 - \lambda^1 q + \lambda^2 q^2 - \lambda^3 q^3 + \cdots)(1 - \lambda^1 q^2 + \lambda^2 q^4 - \lambda^3 q^6 + \cdots) \\
&\quad (1 - \lambda^1 q^3 + \lambda^2 q^6 - \cdots)\cdots
\end{aligned}$$

Collecting terms we find that the first few coefficients are

$$\begin{aligned}
&-\lambda^1 \\
&\lambda^2-\lambda^1 \\
&-\lambda^3+(\lambda^1)^2-\lambda^1=-\lambda^3+\lambda^1(\lambda^1-1) \\
&[\lambda^4-(\lambda^2-\lambda^1)(\lambda^1-1)] \\
&-\lambda^5+\lambda^1(\lambda^3-\lambda^2+\lambda^1)-\lambda^1(\lambda^2-\lambda^1+1) \\
&\lambda^6+\lambda^1(-\lambda^4+\lambda^3-\lambda^2+\lambda^1)+(-\lambda^1)^2(\lambda^1-1)+(\lambda^2)^2 \\
&\vdots
\end{aligned}$$

Note: Because of the factor q, the sum starting with λ^n corresponds to the representation θ_{n+1}. Also note that the expression for θ_6 collapses to $\theta_2\theta_3$.

Analysis of these increasing complicated combinations of exterior powers should lead at least to known properties of the sequence $\tau(n)$ and possibly to quite new results. The first few checks on dimensions give

$$\begin{aligned}
q-24q^2+252q^3-1472q^4+4830q^5-6048q^6-16744q^7+84480q^8 \\
-113643q^9-\cdots-25499225q^{25}\cdots-73279080q^{27}+\cdots
\end{aligned}$$

Note: For small powers of p Recurrence Relation (2) gives $\tau(p^2)=\tau(p)^2-p^{11}$, and $\tau(p^3)=\tau(p)(\tau(p^2)-p^{11})$.

The known congruences satisfied by numbers $\tau(n)$ provide information about dimensions of certain representations of M_{24}. To understand these better, we must determine the structure of the representation ring $R(M_{24})$ as a λ-ring and isolate recurrent phenomena in the λ^*-expressions for the θ_n, when n is a power of a prime. The following may also provide more than a passing interest: Listing irreducible characters $\chi_n (n=1,2,\ldots,26)$ and restricting attention to those taking integral values, find the least value of m such that χ_n contributes to the character of θ_m. By inspection $m(1)=1, m(23)=2$, and $m(253)=3$.

4.6. Notes

Without doubt the best reference for the Mathieu groups is the "*Atlas*" [35], containing as it does, at least in outline, all known constructions of this fascinating family. Perhaps because of our topological background we start with the projective linear group $L_3(4)$ and regard M_{24} as being obtained by extension from it. Character tables of M_{12} and M_{24} in characteristic zero were first constructed by I. Schur, who also observed that M_{12} is contained in the larger group as a subgroup.

The Todd and Coxeter representations are described in two papers in [116, 117]. With the exception of one high dimensional representation of M_{24} in characteristic 2, the modular character theory of the Mathieu family was given by [62], on which paper we have relied heavily. Another source is R. Parker's modular supplement to the main atlas. We first found out about the relation between conjugacy classes of elements in M_{24} and Hecke eigenspaces of modular forms from the paper [81]. Its existence is truly serendipitous and deserves to be better known, particularly by number theorists. Having learned the bare minimum about Ramanujan numbers from the book [95], we learned much more from the three quite different surveys by Rankin himself [9, pp. 245–268], M. Ram Murty [9, pp. 269–288], and H. P. F. Swinnerton-Dyer [9, pp. 289–311] in the Ramanujan centenary volume. Melding arithmetic and a version of the representation theory of M_{24} that takes account of exterior powers should be a very fruitful area for research.

5

Cohomology of Certain Simple Groups

As mentioned in the Introduction, our aim is to calculate and if possible find a geometric model for $Ell^*(BG)$, where G is a finite group. Before doing this it makes sense to look at the ordinary cohomology of BG, since methods used are common to all complex-oriented theories with products. We wish to determine *stable* elements in $H^*(BG_p, \mathbb{Z})$, where G_p runs through a representative family of Sylow subgroups for all (odd) primes dividing the order of G. When G_p is cyclic, the fact that BG_p fibers over $\mathbb{C}P^\infty$, and an application of the Gysin sequence (compare Sec. 3.4), show that $H^*(BG_p, \mathbb{Z})$ is polynomial on a two-dimensional generator. This solves the problem for so-called groups with periodic cohomology, such as $SL_2(p)$, and gives partial information in many more cases. Groups that we have particularly in mind are M_{12}, M_{23}, and M_{24} — indeed much of Chap. 5 is devoted to the last of these. However given the motivational importance of Moonshine for the definition of elliptic objects in Chap. 8, we also include some calculations for the second group Co_2 in the Conway series and for the Monster simple group $\mathbb{M}$. Detailed calculations at the prime 3 for BM_{24} are due to [49]. With minor modifications, we follow his argument.

5.1. $H^*(BM_{24}, \mathbb{Z})_{(\mathrm{odd})}$

Let G be a finite group. For the topologist the integral cohomology ring $H^*(G, \mathbb{Z})$, or more generally $H^*(G, A)$ with G acting trivially on the coefficients A, is that of the classifying space $BG = K(G, 1)$. Up to homotopy the connected space $K(G, 1)$ is characterized as having $\pi_1(K, *) \cong G$ and $\pi_i(K, *) = 0$ $(i \geqslant 2)$. A cellular model for BG is obtained by starting with a two-dimensional complex $K^{(2)}$ with fundamental group equal to G, generators of a presentation correspond to the 1-cells attached to a base point $*$, the relations to 2-cells. We then attach cells inductively to eliminate higher homotopy groups. (Algebraically use the free differential calculus to construct $C_*(K^{(2)})$, then take kernels to obtain a projective resolution of $\mathbb{Z}$.)

EXAMPLE. Compare Sec. 3.4: the cyclic group C_{p^t} has classifying space $\bigcup_{n=1}^{\infty} S^{2n-1}/C_{p^t}$, where C_{p^t} (with generator A) acts by means of a direct sum of copies of the representation $A \mapsto e^{2\pi i/p^t}$.

The general definition of Chern classes in Chap. 1 can be adapted to define Chern classes of a representation $\rho : G \to U_n$. Thus if EG denotes the universal covering space of BG, with G identified with the deck transformation group, acting on the right as in Sec. 3.1, we can construct the flat complex vector bundle:

$$\begin{array}{ccc} \mathbb{C}^n & \longrightarrow & EG \underset{G}{\times} \mathbb{C}^n \\ & & \downarrow \\ & & BG. \end{array}$$

DEFINITION.

1. $c_k(\rho) = c_k(EG \underset{G}{\times} \mathbb{C}^n)$
2. The Chern subring $Ch(G) \subseteq H^{\text{even}}(BG, \mathbb{Z})$

is the subring of the even-dimensional cohomology generated by Chern classes of the irreducible representations $\rho_1 \cdots \rho_s$ ($s =$ number of conjugacy classes in G).

In Chap. 3, we mentioned Mackey functors, which require both restriction and induction maps. Axioms for Chern classes (see Sec. 1.1) imply that these are compatible with restriction; determining the Chern classes of an induced representation is more subtle. Recall the recursive definition of Newton polynomials by the formula:

$$s_k - c_1 s_{k-1} + c_2 s_{k-2} + \cdots + (-1)^k k c_k = 0,$$

that is

$$\begin{aligned} s_1 &= c_1 \\ s_2 &= c_1^2 - 2c_2, \end{aligned}$$

and so forth.

(The polynomial s_k expresses the symmetric function $x_1^k + x_2^k + \cdots$ in terms of elementary symmetric functions.) We then have the following very useful induction property: Let the order of G be p^n ($p \geqslant 3$) and let ρ be a one-dimensional representation of the index p normal subgroup K of G, then:

$$\begin{aligned} s_k(i_! \rho) &= i_*(s_k \rho) = i_*(c_1 \rho)^k, \qquad k < p-1, \quad k = p, \\ s_{p-1}(i_! \rho) &= i_*(s_{p-1} \rho) + (p-1) j^*(\alpha^{p-1}), \end{aligned}$$

where α generates $H^2(G/K,\mathbb{Z})$ and $j : G \to G/K$ is the quotient homomorphism.

For small values of k this relation is a consequence of Galois invariance; for $k = (p-1)$ we must allow a correction term. For more details we refer the reader to [113, Theorems 6.3 and 8.7]. The following exercise shows the need for a correction term:

EXERCISE. Evaluate Chern classes of $i_*(1)$ when $K = \{1\}$ and $G = C_p$.

For spaces BG the single most important fact about their cohomology is that the restriction map:

$$i^* : H^*(BG,\mathbb{Z})_{(p)} \to H^*(BG_{(p)},\mathbb{Z})$$

from the p-torsion subgroup of the cohomology of BG to that of BG_p is injective. The image consists of stable elements, i.e., those elements in $H^*(BG_p,\mathbb{Z})$ that have the same image in $H^*(B(G_p \cap gG_pg^{-1}),\mathbb{Z})$ as their conjugates in $H^*(B(gG_pg^{-1}),\mathbb{Z})$. Under favorable circumstances, for example when G_p is abelian, stable elements coincide with those invariant under the action of the normalizer $N(G_p)$ modulo the centralizer $C(G_p)$. These results continue to hold in the more general setting in Chap. 6.

If we let p be an odd prime and G a group such that a p-Sylow subgroup G_p is cyclic, we obtain the following useful result in Theorem 5.1.

THEOREM 5.1. *Let G_p be cyclic. There exists a virtual representation ρ of G such that the restriction of the top dimensional Chern class $c_d(\rho)$ generates the image of $H^{2d}(G,\mathbb{Z})$ in $H^{2d}(G_p,\mathbb{Z})$. Furthermore taking cup products with this class defines an isomorphism:*

$$H^i(G,\mathbb{Z})_{(p)} \xrightarrow[-]{\sim} H^{i+2d}(G,\mathbb{Z})_{(p)}$$

for all $i \geqslant 1$.

Proof: See [113]. It is well-known (see the preceding Example) that $H^*(BG_p,\mathbb{Z})$ is a polynomial algebra over $\mathbb{Z}/p^t$ on a two-dimensional generator, which can be taken as the first Chern class of the one-dimensional representation used to define BG_p. ■

The representation ρ in Theorem 5.1 belongs to the preimage of a d-dimensional representation σ of $N(G_p)$ obtained by first extending a one-dimensional representation of G_p trivially to $C(G_p)$, then inducing to the normalizer. Note: We have also shown that $d = [N : C]$.

REMARKS.

1. A group G such that G_p is cyclic is said to be p-periodic (in cohomology). The integer $2d$ is called the p-period.
2. If $p = 2$ we can define a 2-period if G_2 is cyclic ($d = 1$) or generalized quaternionic ($d = 2$).

The Mathieu group M_{24} has order $2^{10} \cdot 3^3 \cdot 5 \cdot 7 \cdot 11 \cdot 23$, and is periodic for all primes greater than 3. Inspecting the character table in [35] shows that we have periods:

p	5	7	11	23
$2d_p$	8	6	20	22.

PROPOSITION 5.1. *If T denotes the Todd representation of M_{24} in $GL_{11}(\mathbb{F}_2)$ with partial character given in Sec. 4.1, then $c_{d_p}(T)$ generates the p-torsion in integral cohomology for $p \geqslant 5$.*

Proof: This is by direct calculation. Values of the Todd character on cycle types that split over M_{24} are explained by the formula:

$$\frac{-1+i\sqrt{n}}{2} = \frac{1}{2}\sum_{j=1}^{n-1} z^{j^2} (n \equiv -1(4)), \qquad z = e^{2\pi i/n}.$$

$Aut(M_{24,p})$ has order $(p-1)$, and the image of the normalizer of a p-Sylow subgroup (modulo its centralizer) has index 2 or 1, depending on whether an S_{24}-conjugacy class splits or not.

We complete the calculation for the prime 7: if σ denotes the representation mapping the generator $1^3 7^3$ of the positive copy of $M_{24,7}$ to a primitive seventh root of unity, then the restriction of T decomposes as:

$$(\rho_{reg}) + (\sigma + \sigma^4 + \sigma^2) + (1).$$

Therefore:

$$c.(T)_{(7)} = (1-s^6)(1+s)(1+2s)(1+4s),$$

where $s = c_1(\sigma)$. Expanding this expression:

$$c_3(T)_{(7)} = s^3,$$

providing the required generator is in dimension 6. ■

The prime 3 is much more interesting. A 3-Sylow subgroup has order 27, where for a general odd prime number p, we define the extra-special group

$$p_+^{1+2} = \langle A, B, C | A^p = B^p = C^p = 1,\ C = [A,B],\ [C,A] = [C,B] = 1 \rangle.$$

THEOREM 5.2. *The cohomology ring $H^*(B(p_+^{1+2}),\mathbb{Z})$ is generated by elements in the following table:*

Generator	α_1,α_2	ν_1,ν_2	κ_i	ζ
Degree	2	3	$2i$	$2p$
			$2\leqslant i\leqslant p-1$	
Additive order	p	p	p	p^2.

In particular the even-dimensional cohomology is generated by Chern classes.

Proof: This falls into two parts; first we describe the Chern subring, then we use the spectral sequence of an extension to show that we have a complete set of generators for H^{even}. The two generators in degree 3 give us no trouble; indeed in the analogous argument for more general cohomology theories, they disappear.

The irreducible representations of (p_+^{1+2}) are either one-dimensional and factor through the commutator quotient group generated by classes of A and B or p-dimensional. A representation ξ of the latter type is obtained by induction from the normal subgroup $\langle B,C\rangle$; and (with $\hat{\gamma}=e^{2\pi i/p}$) it has the matrix form:

$$A\longmapsto\begin{pmatrix}0&1&\cdots&0\\0&0&1\cdots&0\\\vdots&&&\\1&0&\cdots&0\end{pmatrix},$$

$$B\mapsto \operatorname{diag}(1,\hat{\gamma},\ldots,\hat{\gamma}^{p-1})$$
$$C\mapsto \operatorname{diag}(\hat{\gamma},\ldots,\hat{\gamma}).$$

Applying the induction formula for Chern classes, we obtain generators α_1,α_2 (from one-dimensional representations) and $\kappa_j=s_j(\xi), j=2,\ldots,p-1$, and $\zeta=c_p(\xi)$ in dimension $2p$.

Generators κ_j are obtained by cohomological induction with a correction term in degree $p-1$. This correction term is detected in relations between generators, to which we refer as little as possible. The order of each generator equals p except for $\zeta=c_p(\xi)$, which equals p^2. This follows because the expansion of the Newton polynomial $s_p(\xi)$ contains $p\zeta$. Note: In this connection the exponent of $H^*((p_+^{1+2}),\mathbb{Z})$ is strictly greater than that of the group itself.

It remains to show that there are no additional generators of even degree. With $G=(p_+^{1+2})$ consider the spectral sequence of the central extension:

$$1\to C_p^C\to G\to C_p^{\overline{A}}\times C_p^{\overline{B}}\to 1$$

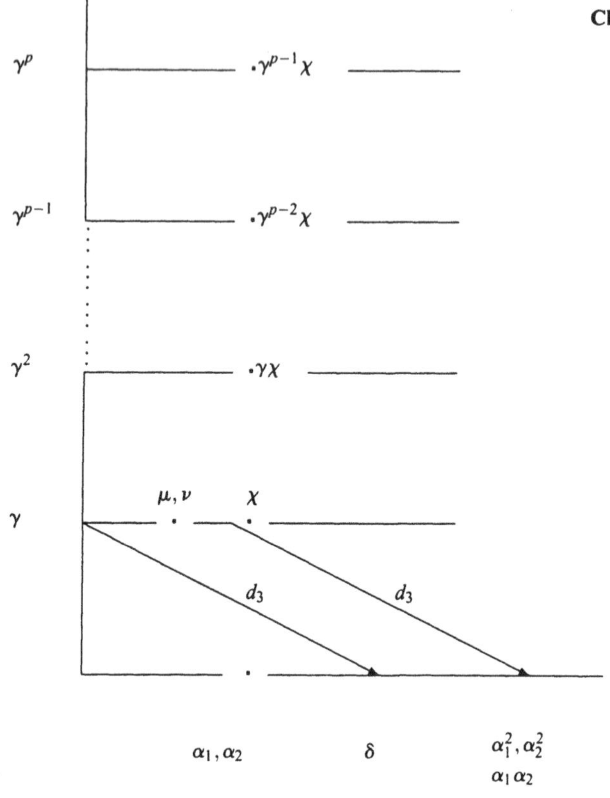

FIGURE 5.1. E_2-page of the spectral sequence (H^*).

with

$$E_2^{i,j} \cong H^i(C_p \times C_p, H^j(C_p, \mathbb{Z})).$$

Since $H^{\text{odd}}(C_p, \mathbb{Z}) = 0$, the even-dimensional cohomology is detected by subquotients of $E_2^{2i,2j}$; since the extension is central:

$$E_2^{0,2j} \cong H^{2j}(C_p, \mathbb{Z}),$$

generated by γ^j where γ is the first Chern class of the representation mapping C to $\hat{\gamma}$ just given. Note: γ cannot survive as an infinite cycle, and it has a nonzero image under d_3. But the vertical multiplication:

$$\cup\gamma : E_2^{i,j} \longrightarrow E_2^{i,j+2}$$

is a monomorphism for $j \geqslant 0$ (an isomorphism for $j > 0$), and the spectral sequence admits a horizontal multiplication:

$$E_2^{i,2j} \circ E_2^{k,2j} \longrightarrow E_2^{i+k,2j}, \qquad (j>0),$$

defined by the product in the cohomology of $C_p \times C_p$ for fixed coefficients. In Fig. 5.1, which describes page 2 near the origin, $\mu, \nu \in E_2^{1,2}$ are independent generators and $\chi = \mu \circ \nu$ in $E_2^{2,2}$. On later pages of the spectral sequence, the vertical isomorphism $\cup\gamma$ is replaced by $\cup\chi\gamma$. This follows since: $H^2(G,\mathbb{Z})$ is of type $(p,p), d_3\gamma = s\delta, s \equiv 1(p)$, so:

$$d_3\gamma^i = si\gamma^{i-1} \cdot \delta \neq 0 \qquad (1 \leqslant i \leqslant p)$$

by vertical periodicity. The generators of $H^4(C_p \times C_p, \mathbb{Z})$ are linearly independent, so $d_3\mu = d_3\nu = 0$; since μ, ν come from the exterior subalgebra of $H^*(C_p \times C_p, \mathbb{F}_p)$ their powers do not contribute to $H^{\text{even}}(G,\mathbb{Z})$. The elements $\gamma^i\chi$ must all survive to infinity $(0 \leqslant i \leqslant p-3)$, since they are needed to detect summands of order p corresponding to Newton polynomials:

$$s_{i+1}(\xi) = \operatorname{Ind}(\gamma^{i+1}), \qquad 0 \leqslant i \leqslant p-3.$$

This almost holds in dimension $2p$ except that we must work with the pair $(\gamma^p, \gamma^{p-2}\chi)$ on the line of total degree $2p$ to obtain the correct order for the top dimensional Chern class. By an explicit resolution it is possible to check the preceding assertion that H^3 is generated by the independent generators μ, ν and also that H^4 is generated by $\alpha_1^2, \alpha_1\alpha_2, \alpha_2^2$ and χ.

By restricting the top dimensional Chern class to the subgroup $\langle B, C\rangle$, we see that the fiber term $\gamma^p \neq 0$. Then taking vertical products defines an isomorphism:

$$\gamma^p : E_2^{i,j} \to E_2^{i,j+2p} \qquad j > 0.$$

Up to periodicity and known products, we have now described the E_2-page of the spectral sequence; the E_3-page is identical, since odd rows are zero.

The differential d_3 can be explicitly described in low dimensions (see the preceding), and product structures dictate its behavior otherwise. The effect is to introduce relations between generators, such as $\alpha_1\mu = \alpha_2\nu$. Although we have now detected enough elements to generate $H^*(G,\mathbb{Z})$ it actually follows that in degrees $\leqslant 2p, E_4 = E_\infty$.

The spectral sequence (Fig. 5.1) is explained in exhaustive detail in [76], and we have done little more than improve the detection of universal cycles using Chern classes. It seems clear that Lewis's most critical observations are the universality of the classes $\gamma^i\chi (0 \leqslant i \leqslant p-2)$ [76, Lemma 6.18] and γ^p [76, Lemma 6.9]. He also gives an explicit proof, independent of characteristic classes, of the nontriviality of cohomological induction from $\langle B, C\rangle$ to G [76, Lemma 6.22]). ■

REMARK. The sketch proof just given hides an ambiguity in the notation in the literature. Lewis defines the generator κ_{p-1} as $\mathrm{Ind}(\gamma^{p-1})+\alpha_2^{p-1}$, while other authors (including Green) prefer $s_{p-1}(\xi)$. The induction formula Sec. 5.1 shows that:

$$\kappa_{p-1}^{\text{Lewis}} = \kappa_{p-1}^{\text{Green}} + \alpha_1^{p-1} + \alpha_2^{p-1}.$$

Because of our use of characteristic classes, we prefer the second notation. The alert reader will notice the change only in the use of certain multiplicative relations.

We now apply Theorem 5.2 to the group M_{24}, noting that since $p=3$, there is only one generator of type κ_i, namely, $\kappa=\kappa_2$ of degree 4.

THEOREM 5.3. *The ring $H^*(M_{24},\mathbb{Z})_{(3)}$ has generators as in the following table:*

Generator	β	θ	ν	ξ
Degree	4	16	11	12
Additive order	3	3	3	9,

These are subject to relations $\nu^2=0, \beta\theta=0$. Chern classes of the Todd representation T generate the even-dimensional subring.

Proof: Up to easy manipulations this is given in [49], using Lewis's relations modified by the preceding choice of κ_{p-1}.

STEP 1: 3-local structure. Write $P=M_{24,3}\cong 3_+^{1+2}$ for a representative 3-Sylow subgroup with presentation in terms of generators A,B,C given earlier. We have a short exact sequence:

$$P \rightarrowtail N_{M_{24}}(P) \leftrightarrows D_8$$

where the dihedral group D_8 acts on P by its inclusion (unique up to conjugacy) in the semidihedral subgroup SD_{16} in $GL_2(3)\cong Out(P)$. The group D_8 has generators J and K, which as automorphisms act as follows:

$$J\begin{cases} A\longmapsto B^{-1} \\ B\longmapsto A \\ C\longmapsto C \end{cases} \qquad K\begin{cases} A\longmapsto B^{-1} \\ B\longmapsto A^{-1} \\ C\longmapsto C^{-1} \end{cases}$$

The Mathieu group contains two conjugacy classes of elements of order 3, one contained in $1^6 3^6$, the other in 3^8. In terms of elements of P one of these contains $\{A^rC^t, B^rC^t; r=1,2, t=0,1,2\}$; the other contains $\{A^rB^{\pm r}C^t; r=1,2, t=0,1,2\}$.

STEP 2: $H^*(N,\mathbb{Z})_{(3)}$ is generated by $\alpha = \alpha_1^2 + \alpha_2^2$, κ, $\eta = \zeta^2$, $\nu = (\alpha_1\nu_1 + \alpha_2\nu_2)\zeta$ subject to $\nu^2 = 0$, $\alpha^2 = \kappa^2$.

Proof: Having identified the generators in $H^*(M_{24,3},\mathbb{Z})$ it is easy to describe the induced action of the automorphisms J and K, thus:

$$J^* \begin{cases} \alpha_1 \mapsto \alpha_2 \\ \alpha_2 \mapsto -\alpha_1 \\ \zeta \text{ is fixed} \\ \kappa \text{ is fixed} \end{cases} \qquad K^* \begin{cases} \alpha_1 \mapsto -\alpha_2 \\ \alpha_2 \mapsto -\alpha_1 \\ \zeta \mapsto -\zeta \\ \kappa \text{ is fixed} \end{cases}$$

In odd degrees νs follow αs under J^*, and also under K^* (up to change of sign). Working over $\mathbb{F}_9$ rather than $\mathbb{F}_3$, i.e., adjoining $i = \sqrt{-1}$ to the coefficients, we see that:

$$J^*(\alpha_1 - i\alpha_2) = i\alpha_1 + \alpha_2,$$
$$J^*(\alpha_1 + i\alpha_2) = -i\alpha_1 + \alpha_2,$$

so that J^* fixes $\alpha_1^2 + \alpha_2^2$ and

$$(\alpha_1 \pm i\alpha_2)^4 = \alpha_1^4 + \alpha_2^4 = -\kappa(\alpha_1^2 + \alpha_2^2).$$

Here we use the relations $\alpha_1\alpha_2^3 = \alpha_1^3\alpha_2$ [76, Lemma 6.20] and $\alpha_i\kappa = -\alpha_i^3$ [76, Lemma 6.25]; the latter is a variant of the expression for $s_2(i_!\xi)$ in the induction formula.

Similar reasoning shows that J^* fixes $\mu = \alpha_1\nu_1 + \alpha_2\nu_2$ in degree 5. The induced automorphism K^* fixes κ and α (obviously) and multiplies both ζ and μ by -1. The relation $\kappa^2 = \alpha^2 = \alpha_1^4 - \alpha_1^2\alpha_2^2 + \alpha_2^4$ is again implicit in those given by [76].

STEP 3: $Ch(M_{24})$ contains $\beta = \alpha + \kappa$, $\xi = \eta - \kappa^3$, and $\theta = (\alpha - \kappa)\eta$.

Proof: On lifting T to characteristic zero and restricting to the conjugacy classes 1^{24}, $1^6 3^6$, and 3^8 we obtain a sum of 5 one-dimensional and 2 three-dimensional irreducible representations of p_+^{1+2}. The missing one-dimensional representations are those mapping both generators A and B nontrivially, (exercise). Hence with a slight abuse of notation and recalling the definition of κ and ξ in terms of Chern classes:

$$\begin{aligned} c\cdot(T) &= (1-\alpha_1^2)(1-\alpha_2^2)(1+\kappa+\zeta)(1+\kappa-\zeta) \\ &= (1-\alpha_1^2)(1-\alpha_2^2)(1-\kappa+\kappa^2-\eta) \\ &= 1-(\alpha+\kappa)-(\eta-\kappa^3)+(\alpha\eta-\alpha\kappa^3-\kappa^4)+(\alpha\kappa+\kappa^2)\eta. \end{aligned}$$

Note: Terms of degree 8 cancel, and we remove factors $\alpha_1^2\alpha_2^2$ by means of relations $\alpha_i\kappa = -\alpha_i^3$ and $\kappa^2 = \alpha_1^4 - \alpha_1^2\alpha_2^2 + \alpha_2^4$. Relabeling with $\beta = \alpha + \kappa, \xi = \eta - \kappa^3$ (ξ is now an element in cohomology, *not* a representation) and $\theta = (\alpha - \kappa)\eta$ we have

$$c_8(T) = -(\theta + \beta\xi + \beta^4).$$

The argument so far has obtained bounds for $H^*(BM_{24}, \mathbb{Z})$, and it remains to show that a stable element can be expressed in terms of ν (from Step 2) and β, ξ, θ (from Step 3).

STEP 4: The lower bound of $H^{\text{even}}(BM_{24}, \mathbb{Z})_{(3)}$ suffices.

Proof: The computational trick, which is really an exercise in linear algebra, is to approximate the action of an arbitrary automorphism of P on its cohomology by one that further induces an action on $H^*(N, \mathbb{Z})_{(3)}$. Eliminating trivial cases, we are reduced to considering conjugates P^g of P such that $P \cap P^g = \langle D, C\rangle$, with C central in P and D central in the conjugate subgroup P^g. Note: D is not central in P. The computational trick allows us to construct an automorphism Ψ of P such that $D \mapsto C^g, C \mapsto D^g$ and Ψ^* fixes elements in $H^*(N, \mathbb{Z})_{(3)}$. Since the following diagram commutes

$$\begin{array}{ccc} P^g & \longrightarrow P \xleftarrow{\Psi} & P \\ i\uparrow & & \uparrow i' \\ \langle D, C\rangle & \underset{\Phi=\text{switch}}{\longrightarrow} & \langle C, D\rangle, \end{array}$$

after restriction to the cohomology of $\langle D, C\rangle$, stable elements are those fixed by Φ^*.

Since we already know that Chern classes are stable, the stability of $(\alpha + \kappa)\eta$ and hence of $\kappa\eta$ follow from the relation:

$$(\alpha + \kappa)(\eta - \kappa^3) - (\alpha + \kappa)^4 = (\alpha + \kappa)\eta.$$

Now take an arbitrary stable element of even degree and

1. Subtract powers of $\eta - \kappa^3$ to remove lone powers of η.
2. Remove lone powers of κ using the relation $(\alpha + \kappa)^{t+1} = (-1)^t\kappa^t(\alpha + \kappa)$.
3. Appeal to Φ^*-invariance to eliminate lone monomials $\alpha\kappa^t$.

We are left with an expression divisible by $\alpha\eta$ or $\kappa\eta$. Use similar manipulations to the preceding and the relation $\alpha^2 = \kappa^2$ to reduce ξ to $\kappa\eta\xi'$. Since $\kappa\eta$ is stable, we can use induction to show that ξ has to be expressible in terms of the given Chern classes. In odd dimensions the stability of $\nu = \alpha_1 \nu_1 + \alpha_2 \nu_2$ is clear. We are done. ■

The proof of Theorem 5.3 suggests the following remark.

REMARK. The modular representation ring $R\overline{\mathbb{F}}_2(M_{24})$ is generated as a λ-ring with conjugation by the class of T.

We label this as a remark rather than a corollary, since the cohomological calculation implies only that the I-adic completion is generated by T. To check the stronger statement we use the calculations in [62], for example:

$$44 = \lambda^2 T - T, \qquad 120 = T.\overline{T} - 1, \ldots$$

5.2. Remaining Mathieu Groups

Associated with the chain of Steiner systems discussed in Sec. 4.1 is the chain of groups:

$$PSL_3(\mathbb{F}_4) = M_{21} \hookrightarrow M_{22} \hookrightarrow M_{23} \hookrightarrow M_{24}.$$

Although it is not sporadic the projective special linear group M_{21} carries much of the structure of $H^*(BM_k, \mathbb{Z})_{(3)}$ for $k = 22$ and 23, since in these cases $M_{k,3}$ is elementary abelian of rank 2. We have

1. $$C(M_{k,3}) = M_{k,3} \qquad (k = 21, 22, 23).$$

2. $$\begin{aligned} N(M_{k,3})/C(M_{k,3}) &\cong Q_8 &&\text{(quaternion group,}k = 21, 22),\\ &\cong SD_{16} &&\text{(semidihedral group,}k = 23). \end{aligned}$$

The group SD_{16} has presentation:

$$\{R, T | R^8 = T^2 = 1, T^{-1}RT = R^3\}.$$

Isomorphisms can be obtained from the *Atlas* [35] (compare Step 1 in the proof of Theorem 5.3). Since the centralizer is as small as possible the action of the quotient group on $M_{k,3}$ is faithful. When $k = 21$, we write G for the normalizer; it is a split extension of the form:

$$C_3^A \times C_3^B \rightarrowtail G \overset{\leftarrow}{\twoheadrightarrow} Q_8^{S,T}.$$

We select a convenient basis for the normal subgroup as a vector space over an extension field of $\mathbb{F}_3$. Generators A and B as usual correspond to $\alpha, \beta \in H^2(C_3 \times C_3, \mathbb{Z})$.

Outside of the primes 2 and 3 calculations similar to those in Proposition 5.1 yield Proposition 5.2.

PROPOSITION 5.2. *If $k = 22$ or 23 and $p = 5, 7$, or 23, $H^*(BM_k, \mathbb{Z})_{(p)}$ is generated by the restriction of $c_j(T)$, $j = 4, 3$, or 11 respectively. At $p = 11$, a symmetric conjugacy class splits, and $c_{10}(T)$ must be replaced by the Chern class of a 280-dimensional ($k = 22$) or 896-dimensional representation ($k = 23$).*

Now let us consider the prime 3.

THEOREM 5.4. *If $k = 22$ or 23, the subring $Ch(M_k)_{(3)}$ is generated by $c_j(T|M_{k,3})$, $j = 3, 4$. In both cases the Chern subring is properly contained in $H^*(BM_k, \mathbb{Z})_{(3)}$.*

Proof: We calculate the 3-primary part of $H^*(BG, \mathbb{Z})$, where G is the normalizer of a representative 3-Sylow subgroup in $PSL(3, \mathbb{F}_4)$. The spectral sequence for the defining short exact sequence is trivial, so $H^*(BG, \mathbb{Z})_{(3)} = H^*(B(C_3 \times C_3), \mathbb{Z})^{Q_8} = E_2^{*,0}$. The odd-dimensional contribution is an exterior algebra on a three-dimensional generator (compare [76]). In even dimensions we proceed as follows: Let V be a two-dimensional vector space over $\mathbb{F}_3$ and consider the induced action of Q_8 on the symmetric algebra $Sym(V^*)$. Take coefficients in $\mathbb{F}_9$ rather than $\mathbb{F}_3$ to diagonalize the action of an element S of order 4 in Q_8. Here we use the usual presentation of Q_8 as:

$$\{S, T | S^4 = 1, \quad S^2 = T^2, \quad T^{-1}ST = S^{-1}\}.$$

We represent Q_8 in $SL_2(\mathbb{F}_9)$ by:

$$S \longmapsto \begin{bmatrix} i & 0 \\ 0 & -i \end{bmatrix} T \longmapsto \begin{bmatrix} 0 & -1 \\ 1 & 0 \end{bmatrix}.$$

Having extended the scalars, we choose a basis of eigenvectors $\{x, y\}$ for $\mathbb{F}_9 \underset{\mathbb{F}_3}{\otimes} Sym(V^*) = \mathbb{F}_9[\alpha, \beta]$, with $Sx = ix$ and $Sy = -iy$. Formally we first choose x, then take y to be the image of x under the Frobenius map. As an automorphism Fr fixes α and β, and on the coefficients $Fr(\lambda) = \lambda^3$. We further suppose that over the extension field $\mathbb{F}_9$ bases $\{x, y\}$ and $\{\alpha, \beta\}$ are related by the equations:

$$x = i\alpha + \beta, \qquad y = \alpha + i\beta.$$

The remark about the choice of basis is now clear — G_3 is generated by A and B dual to the classes α and β. Now $\mathbb{F}_9[x, y]^{\langle S \rangle}$ has an $\mathbb{F}_9$-basis consisting of all monomials $x^j y^k$ with $j+3k \equiv 0 \pmod 4$. This is equivalent to $(k-j) \equiv 0 \pmod 4$. Since the action of the generator T induces the automorphism $x \longmapsto -y, \quad y \longmapsto x$, one type of invariant polynomial is evenly symmetric in x and y, i.e., we consider symmetric polynomials of the form:

$$x^j y^k + x^k y^j = \sigma_{jk}^+,$$

where j and k are both even and $(k-j) \equiv 0 \pmod 4$. The second type must satisfy $x^j y^k - x^k y^j = \sigma_{jk}^-$, where j and k are both odd and $(k-j) \equiv 0 \pmod 4$. The first few invariant polynomials are $x^2y^2 = -(\alpha^2+\beta^2)^2, x^4+y^4 = -(\alpha^4+\beta^4)$, $x^5y - xy^5 = (\alpha^2+\beta^2)(\alpha^3\beta+\alpha\beta^3), \ldots$. We see immediately that $Sym(V^*)^{Q_8}$ has two generators of degree 4, one of degree 6, $\cdots$. On the other hand by counting dimensions, we see that all but one of the irreducible representations of G factor through the quotient group Q_8, and the exception, obtained by induction from the trivial representation, restricts to the regular representation minus a trivial summand on $C_3 \times C_3$. An easy calculation now shows that $Ch(G)_{(3)}$ is generated by c_4 and c_6 of this restriction, and hence it is properly contained in $H^{\text{even}}(G,\mathbb{Z})_{(3)}$.

This argument applies immediately to Mathieu groups M_{21} and M_{22} since stable elements in the cohomology of the abelian groups $M_{k,3}$ coincide with those invariant under the normalizer. This is proved for a general cohomology theory in Chap. 6. Inspecting the character table again shows that $Ch(M_{22})_{(3)}$ is generated by Chern classes of the regular representation of $C_3 \times C_3$. For M_{23} the argument follows the same pattern except that we replace Q_8 by SD_{16}, represented over $\mathbb{F}_9$ by:

$$S \longmapsto \begin{bmatrix} \zeta & 0 \\ 0 & \zeta^3 \end{bmatrix}, \qquad R \longmapsto \begin{bmatrix} 0 & 1 \\ 1 & 0 \end{bmatrix},$$

where $\zeta = 1-i$ is a primitive 8th root of unity. A basis of eigenvectors is given by $\{x,y\}$, where $Sx = \zeta x$, $Sy = \zeta^3 y$; because R has order 2 rather than 4, invariant polynomials are $x^j y^k + x^k y^j$ with $j+3k \equiv 0 \pmod 8$. As we expect this subalgebra is smaller than for M_{22}, but x^2y^2 still provides a generator in degree 4, which is not describable as a Chern class. ■

Proposition 5.2 and Theorem 5.4 show that $H^i(BM_{23},\mathbb{Z})_{(\text{odd})} = 0$ for $i \leqslant 4$. The same holds at the prime 2, although the argument is too delicate to give here. We note however that M_{23} is an example of a group with trivial homology in dimensions less than or equal to 4. At the present state of our knowledge, the existence of such groups is significant for our discussion of elliptic objects in Chap. 8. See especially the assumptions made on the finite group G in Sec. 8.5.

Having discussed the larger Mathieu groups in considerable detail, we summarize more briefly what is known about groups in the chain:

$$(C_3 \times C_3) : Q_8 = M_9 \hookrightarrow M_{10} \hookrightarrow M_{11} \hookrightarrow M_{12}.$$

Corresponding to the Todd representation T, we have a *projective* representation of M_{12} in $PGL_5(\mathbb{F}_3)$, which in contrast to T does not factor through a linear representation. However the mod 3 character table of the point stabilizer subgroup M_{11} does contain an irreducible character C of degree 5; its values on elements of the indicated order are given by:

$$\begin{array}{cccccccc} 1 & 2 & 4 & 5 & 8 & 8 & 11 & 11 \\ 5 & 1 & -1 & 0 & -1+2i & -1-2i & \frac{-1+i\sqrt{11}}{2} & \frac{-1-i\sqrt{11}}{2}. \end{array}$$

The remaining irreducible 3-modular representations can be labeled $1, \bar{5}, 10', 10, \overline{10}, 24$, and 45 [62], with $10'$ taking values:

$$10 \quad 2 \quad 2 \quad 0 \quad 0 \quad 0 \quad -1 \quad -1.$$

PROPOSITION 5.3.

1. *As a λ-ring with conjugation $R\overline{\mathbb{F}}_3(M_{11})$ is generated by the classes of* (5) *and* $(10')$.
2. *$H^*(BM_{11},\mathbb{Z})_{(5)}$ [respectively $H^*(BM_{11},\mathbb{Z})_{(11)}$] is generated by $c_4(C)$ [respectively by $c_5(C)$].*

The proof is by straightforward calculation. Part 2 of Proposition 5.3 above also holds for M_{12}, with Representation C replaced by a virtual representation $(16-10)$ having the same values on the classes $(1^2 5^2)$ and $(1\ 11)$.

These calculations suggest that while the natural characteristic of M_{24} equals 2, that for M_{12} equals 3. At the prime 2 we obtain Theorem 5.5.

THEOREM 5.5.

1. $H^*(BM_{11},\mathbb{F}_2) \cong \mathbb{F}_2[v_3,v_4,v_5]/v_3^2 v_4 + v_5^2 = 0.$
2. *$H^*(BM_{12},\mathbb{F}_2)$ is free and finitely generated over a polynomial subalgebra* $\mathbb{F}_2[d_4,d_6,d_7]$.

Proof: The reader is referred to [8, Chap. 8]. However some remarks are in order. The generators v_i in Part 1 can be expressed in terms of elements d_2 and d_3 by taking $v_3 = d_3$, $v_4 = d_2^2$ and adjoining $d_2 d_3$ as a free generator. Note: $v_3^2 v_4 = -d_3^2 d_2^2$ over $\mathbb{F}_2$. Lower case letter d stands for the Dickson invariant; a major step in the calculations is determining invariant elements $\mathbb{F}_2[a_1,a_2]^{GL_2(\mathbb{F}_2)}$ and $\mathbb{F}_2[a_1,a_2,a_3]^{GL_3(\mathbb{F}_2)}$. (Theorem 5.4 above already includes much of the latter, with $\mathbb{F}_9$ replacing $\mathbb{F}_2$). ■

EXERCISE. Using the character tables in [62] and [8, Theorems 8.1.2 and 8.3.20] determine the subrings in the mod 2 cohomology generated by Stiefel–Whitney classes of real representations for M_{11} and M_{12}.

At the prime 3 M_{12} is best considered as a subgroup of M_{24}. Inspecting the structure of M_{24} in [35] shows a maximal subgroup isomorphic to $Aut(M_{12})$, containing copies of both $3^{1+2}_{+} : D_8$ and $3^2 : GL_2(\mathbb{F}_3)$. Green's methods [49] then exhibit an isomorphism:

$$H^*(BM_{24},\mathbb{Z})_{(3)} \xrightarrow[i^*]{} H^*(BAut(M_{12}),\mathbb{Z})_{(3)},$$

that remains valid when coefficients are reduced mod 3. Changing to the language of Milgram, we see that $H^*(BM_{12},\mathbb{F}_3)$ is free and finitely generated over $\mathbb{F}_3[x_{12},x_{16}]$, with x_{12} and x_{16} corresponding to generators ξ and θ in Theorem 5.3.

5.3. Groups Co_2 and $\mathbb{M}$

Ultimately we hope to understand the cohomology of each member of the happy family of sporadic simple groups, as well as we understand that of the Mathieu series. We illustrate what was already achieved using the second Conway group and the Monster; the importance of the latter is self-evident.

Recall that the group Co_0 is the automorphism group of a certain 24-dimensional (Leech) lattice and the subgroup Co_2 stabilizes a vector v of *type 2*; thus the inner product $\langle v,v\rangle = 4$. The order of Co_2 equals $2^{18}\,3^6\,5^3\,7\;11\;23$, so our methods apply except for the primes 2 and 3. Two results already in the literature are useful — the determination of the 2-modular characters in [107] and of the $\mathbb{Z}$-cohomology at the prime 5 in [111].

THEOREM 5.6.

1. *$H^*(Co_2,\mathbb{Z})_{(7)}$ [respectively $H^*(Co_2,\mathbb{Z})_{(11)}$] is generated by $c_6(22)$ [respectively by $c_{10}(22)$], and $H^*(Co_2,\mathbb{Z})_{(23)}$ is generated by $c_{11}(748)$.*

2. *$H^{even}(Co_2,\mathbb{Z})_{(5)} = Ch(Co_2)_{(5)}$ and has generators:*

Generator	κ	θ_2	θ_3	ξ
Degree	8	24	16	40
Additive order	5	5	5	25

subject to the relations $\kappa\theta_i = \theta_j\theta_i = 0$.

Proof: Part 1 follows by inspecting the character table in [35].

The representation $\lambda^3(22)$ of degree $1540 = 2(22+748)$ decomposes over the prime 2 into irreducibles, one of which is needed to give a generator at the prime 23, since the period is 22 rather than 44.

At the prime 5 we calculate elements in $H^*(5_+^{1+2}, \mathbb{Z})$ invariant under the action of the normalizer of the 5-Sylow subgroup in $GL_2(\mathbb{F}_5)$. Inspecting the *Atlas* shows that this is large (isomorphic to a quadruple cover $4S_4$ of the symmetric group) and thus of index 5 in $GL_2(\mathbb{F}_5)$. Calculating invariants, similar to that already done for M_{22} at the prime 3, shows that $H^*(5_+^{1+2}, \mathbb{Z})^{4S_4}$ is generated by:

$$\kappa, \theta_j = \kappa_j \zeta^{4-j} \quad (j = 2,3) \qquad \xi = \zeta^4.$$

Note: (1) Dickson invariants have degrees 40 and 48 (corresponding to $\kappa^5 + \xi$ and $\kappa\xi$ respectively), entering as invariants for the larger group GL_2, and (2) no combination involving α_1 and α_2 occurs. The full details are contained in the paper [112].

We have now found an upper bound for the 5-primary cohomology; that it is attained follows from a calculation with Chern classes. Integrality and the absence of (α_1, α_2)-components show that it suffices to evaluate the total Chern class of the sum of the four irreducible representations of degree 5. In terms of our standard notation for p_+^{1+2}, we obtain

$$-2\theta_2\zeta^2 + 2\theta_3\zeta + (1+\kappa)^4 - \zeta^4.$$

This in terms of increasing degree expands as:

$$1 - \kappa + (\kappa^2 + 2\theta_3\zeta) - (\kappa^3 + 2\theta_2\zeta^2) + \kappa^4 - \zeta^4.$$

Part 2 of the theorem now follows. ■

For the largest sporadic simple group $\mathbb{M}$, the argument is similar. We omit calculations for $p \geqslant 17$ (cyclic) and list what is known for $p = 11$ (rank 2 and abelian) and $p = 13$ (elementary nonabelian of type 13_+^{1+2}).

- $p = 11$: The even-dimensional cohomology is generated in degrees 40, 60 (2 generators), 80 (2 generators), 160 and 240. This follows by a normalizer (centralizer argument, since $\mathbb{M}_{11}$ is abelian. Note: $N(\mathbb{M}_{11}) = 11^2 : (5 \times SL_2(\mathbb{F}_5))$ with the quotient group contained in $GL_2(\mathbb{F}_{11})$ with index 22. Note also that degrees of Dickson invariants are 220 and 240; we expect the proper Chern subring to be generated by c_{110} and c_{120} for the first nontrivial representation of degree 196883.

- $p = 13$: The calculation is more complicated, the Weyl group $C_3 \times 4S_4$ is generated by a torus T and

$$W = \begin{pmatrix} 0 & 1 \\ -1 & 0 \end{pmatrix}$$

in $GL_2(\mathbb{F}_{13})$, and the stable elements are contained properly in $H^*(13_+^{1+2}, \mathbb{Z})^{\langle T,W\rangle}$. This subring of invariants contains $\{\kappa = \kappa_{12}, \theta_2 \cdots \theta_{11}, \xi = \zeta^{12}\}$; the θ_j are stable, as are Dickson elements $\kappa^{13} + \xi$ and $\kappa\xi$ (of degrees 312 and 336). Dickson elements certainly contribute to the Chern subring. Details are left to the reader (see also [111, Sec. 6]).

The next primes to be considered are $p = 3$ (for Co_2) and $p = 7$ (for $\mathbb{M}$). In both cases it is interesting to note that we are concerned with groups closely related to the group $(p_+^{1+4}) : GL_4(\mathbb{F}_p)$, which is beyond but not too far beyond present investigation techniques.

5.4. Notes

The modern period in the study of the cohomology of finite groups begins with papers by D. Quillen and G. Lewis [93, 76]. Lewis noted in his introduction that before his own calculations, complete or partial information was available for abelian groups, split metacyclic extensions, quaternion and symmetric groups. Quillen's contribution was to show first that calculating $H^*(G, \mathbb{F}_p)$ for G equal to a Chevalley group was much easier than the cohomology of its Sylow subgroups and secondly that "up to nilpotency" it was enough to look at the suitably ordered category of elementary abelian subgroups. The rub is in nilpotency and one of the more important corollaries of Lewis's calculation of $H^*(p_\pm^{1+2}, \mathbb{Z})$ is to show that some cohomology classes cannot be detected on a proper subgroup. Earlier B. Venkov and L. Evens showed that $H^*(G, \mathbb{Z})$ is finitely generated as a module over what we have called the Chern subring, emphasizing the importance of the link with representation theory provided by characteristic classes. As a tool these are exploited systematically in [113].

For an up-to-date introduction to both the general theory and calculation methods, the reader is referred to [8], which contains truly impressive calculations for linear and exceptional Lie groups, as well as for some sporadic simple groups at the prime 2. Several authors have applied variants of Lewis's techniques at odd primes, see in particular the most recent paper on this and the ring of universal stable elements by M. Tezuka and N. Yagita [111]. It is not too far fetched to claim that their methods, those of [8], character tables in [35, 63], and some handy work with computers will go far to unravel $H^*(G, \mathbb{Z})$ for many of the 26 sporadic examples G.

6

$Ell^*(BG)$ — Algebraic Approach

As mentioned in the introduction, a fundamental problem with elliptic cohomology is that however elegant its algebraic definition, we lack a geometric model. In Chaps. 7 and 8 we start to remedy this in the special case when $X = BG$, the classifying space for a finite group. In Chap. 6 we prepare the ground by exploiting the fact that $Ell^*()$ is a complex-orientated cohomology theory with products to reduce the calculation of $Ell^*(BG)$ to a determination of *stable* elements in $Ell^*(BG_p)$. In other words we exploit the formal properties that $Ell^*()$ shares with ordinary cohomology to prove analogs of results in the previous chapter. In particular we obtain a description of $Ell^*(BM_{24})$ in terms of characteristic classes. To do this we need restriction and induction maps, which satisfy Frobenius reciprocity, and a double-coset rule.

In addition if a Sylow subgroup G_p is abelian, stable elements are invariant under the action of the normalizer; if G_p is cyclic, the image of the restriction map can be expressed in terms of a Chern class. A second goal of Chap. 6 is to examine conditions under which $Ell^*(BG)$ is concentrated in even discussions. The importance of this is that $K^{\text{odd}}(BG) = 0$, so it is only reasonable to expect a strict analog of Atiyah's completion theorem for groups satisfying this odd-dimensional condition. However the strongest results in this direction are stated in terms of Morava K-theories rather than elliptic cohomology to exploit the graded field structure of the coefficients. Some of these are described in Sec. 6.3, but also see Appendix A.

6.1. Mackey and Green Functors

Mackey functors are mentioned in Chap. 3 as a necessary tool in axiomatized representation theory. The basic examples are the real and complex representation rings of a finite group, but as the Chap. 5 shows, the concept is much wider.

DEFINITION. Let G be a finite group and let $\mathcal{G}$ be the category whose objects are subgroups H of G and whose morphisms are generated by inclusions $i : K \hookrightarrow H$ and conjugations $c_g : H \to g^{-1}Hg = H^g$ by elements g of G. Let A be a ring. Then a *Mackey functor* $E : \mathcal{G} \rightsquigarrow A$-modules is a family of A-modules $\{E(H) : H \subseteq G\}$ together with maps:

$$\begin{aligned} &i^* = \mathrm{res}_K^H \text{ (restriction)} \\ &i_* = \mathrm{ind}_K^H \text{ (induction or transfer)} \\ &c_g \text{ (conjugation).} \end{aligned}$$

For every $g, h \in G$ and $H, K \in$ objects $(\mathcal{G})$ the following properties hold:

***M*1:** If $L \subset K \subset H$, then $\mathrm{res}_L^K \mathrm{res}_K^H = \mathrm{res}_L^H$ and $\mathrm{ind}_K^H \mathrm{ind}_L^K = \mathrm{ind}_L^H$.

***M*2:** $\mathrm{res}_H^H = \mathrm{ind}_H^H = \text{identity}$.

***M*3:** $c_{gh} = c_h c_g$.

***M*4:** If $h \in H$, then $c_h = \text{identity}$; i.e., $E(H)$ is invariant under inner automorphisms of H.

***M*5:** If $K \subset H$, then $c_g \mathrm{res}_K^H = \mathrm{res}_{K^g}^{H^g} c_g$ and $c_g \mathrm{ind}_K^H = \mathrm{ind}_{K^g}^{H^g} c_g$

***M* = *M*6:** (double-coset formula) $\mathrm{res}_L^H \mathrm{ind}_K^H = \sum_{K\backslash H/L} \mathrm{ind}_{L\cap K^h}^L \mathrm{res}_{L\cap K^h}^{K^h} c_h$.

Here the sum is taken over a set of double-coset representatives $h \in H$.

EXAMPLES. Besides the usual examples from representation theory, we have geometric equivariant bordism. Restriction means restriction of the group action. We define induction by forming the orbit space $G \underset{H}{\times} M$, where H acts on $G \times M$ by $(g,x) \mapsto (gh^{-1}, hx)$ and G acts by left multiplication on G.

EXERCISE. Check the double-coset formula in the previous example.

If in addition the objects $E(H)$ have a multiplicative structure, i.e., $E(H)$ is an A-algebra, we say that E is a *Green functor* if:

***F*1:** All restrictions i^* and conjugations c_g are A-algebra homomorphisms (preserving 1).

***F* = *F*2:** (Frobenius reciprocity)

For $K \subset H$, $a \in E(K)$, $b \in E(H)$,

$$\begin{aligned} \mathrm{ind}_K^H(a \, \mathrm{res}_K^H \, b) &= \mathrm{ind}_K^H(a) \cdot b, \\ \mathrm{ind}_K^H(\mathrm{res}_K^H \, b \, a) &= b \cdot \mathrm{ind}_K^H(a). \end{aligned}$$

Natural transformations of these functors are families of structure preserving maps.

If $F : \mathcal{G} \to A$-algebras is a Green functor, then a *left F-module E* is a Mackey functor $E : \mathcal{G} \to A$-modules and a pairing $F \times E \to E$ such that for $H \subseteq G$, the pairing induces a left $F(H)$-module structure on $E(H)$.

EXERCISE. It is clear that the definition of a Green functor is modelled on that of the character ring of a finite group. The Burnside ring $A(G)$, defined in Sec. 3.6, satisfies the axioms, and it is universal in the sense that every Mackey functor with domain $\mathcal{G}$ is a functor over $\{A(H) : H \subseteq G\}$. For a full proof of this, see [40], although we wish to say more about a special case.

We see shortly that if h^* is a complex-oriented cohomology theory, then:

$$H \mapsto h^*(BH), \qquad H \subseteq G,$$

satisfies axioms for a Mackey functor; in the case of a multiplicative theory, it also satisfies (F). The representability of a cohomology theory easily implies that $h^*(BH)$ is a module over the stable cohomotopy ring $\omega^*(BH)$. More generally we can replace the family of classifying spaces $\{BH : H \subseteq G\}$ by CW-complexes and finite coverings. This is an important point, since many statements about $h^*(BH)$ are proved by first considering a finite approximation $BH^{(n)}$, then choosing h^* to be compatible with limits. That the ω^* action factors through the Burnside ring is shown in [3].

EXERCISE.

1. Define a G-Euler characteristic χ to be a function that assigns to each finite G-complex X an element $\chi(X)$ in some fixed abelian group A. The function χ should satisfy $\chi(\varnothing) = 0$, χ is G-homotopy invariant and (Mayer–Vietoris)

$$\chi(X_1 \cup X_2) + \chi(X_1 \cap X_2) = \chi(X_1) + \chi(X_2).$$

2. Among such functions one is universal. Take A to be the free abelian group with one generator $a_{(H)}$ for each conjugacy class of subgroups (H) in G. Then up to inverting the order of G,A is isomorphic to $A(G)$ — see Lemma 3.2. Given a finite G-complex X, let $r((H),n)$ be the number of G-cells in X of type $(G/K) \times e^n$ with $K \in (H)$, and set

$$\chi(X) = \sum_{(H),n} (-1)^n r((H),n) a_{(H)}.$$

This is the most naive generalization of the usual Euler characteristic; we count G-cells whose sign equals ± 1 depending on the parity of their dimension and keep track of different symmetry types.

3. For the same G-complex, form the bundle fibered by X associated with the universal bundle EG, that is:

$$EG \underset{G}{\times} X \underset{p}{\rightarrow} BG \text{ (compare the definition of Chern classes in 5.1).}$$

Starting with $1 \in \omega^\circ(EG \underset{G}{\times} X)$, or more generally $h^\circ(EG \underset{G}{\times} X)$, we can take the transfer $p_! 1$ [see (4) below], which is again an Euler characteristic.

4. The axioms and the natural action of the class of G/K in (3) show that the characteristic in (2) is indeed universal; i.e., we must have a flat G-set homomorphism $\alpha : A(G) \rightarrow \omega^\circ(BG)$.

6.2. Generalized Group Cohomology

This section is based on [78].

PROPOSITION 6.1. *As H runs through the subgroups of G, $h \mapsto h^*(BH)$ satisfies the conditions for a Green functor.*

Proof: Definition of induction: Let $i : \widetilde{X} \rightarrow X$ be a finite covering of finite CW-complexes and let $l : \widetilde{X} \hookrightarrow \mathbb{R}^L$ be an embedding of $\widetilde{X}$ in some Euclidean space. Set

$$\widetilde{i} : \widetilde{X} \times \mathbb{R}^L \rightarrow X \times \mathbb{R}^L$$

equal to the map:

$$\widetilde{i}(\widetilde{x}, v) = (i(\widetilde{x}), \ell(\widetilde{x}) + v).$$

Then $\widetilde{i}$ defines an embedding of $\widetilde{X} \times D^L$ (with D^L a small disc about 0) into the space $X \times \mathbb{R}^L$ on the right-hand side. The induced map:

$$\widetilde{i}^+ : S^L(X^+) \rightarrow S^L(\widetilde{X}^+)$$

of 1-point compactifications then induces the transfer map i_*.

EXAMPLE. If $\omega^j(X) = \lim_{n\to\infty}[S^n X, S^{n+j}]$, then $i_*(1) \in \omega^\circ(X)$ is represented by the composition:

$$S^L(X^+) \rightarrow S^L(\widetilde{X}) \rightarrow S^L.$$

We can pass to the limit space $BH = \bigcup_n BH^{(n)}$ and ensure that $\varprojlim_n h^*(BH^{(n)}) = h^*(BH)$ because H is finite.

Frobenius reciprocity: This follows from naturality of induction for the diagram of coverings:

$$\begin{array}{ccc} \widetilde{X} & \xrightarrow{(id,i)} & \widetilde{X}\times X \\ {\scriptstyle i}\downarrow & & \downarrow{\scriptstyle i\times id} \\ X & \xrightarrow[\Delta]{} & X\times X, \end{array}$$

where Δ equals the diagonal.

Double-coset rule: Let $X = X(G)$ have finite fundamental group G and $i_K^H : X(K) \to X(H)$ be the finite covering associated with the pair of subgroups $K \subseteq H \subseteq G$. Write $X(1)$ = universal cover of $X(G)$, so that with respect to the (right) action of the deck transformation group $X(G) = X(1)/G$. The claim follows by applying naturality to the Cartesian square of spaces:

$$\begin{array}{ccc} Y & \xrightarrow[k]{} & X(K) \\ {\scriptstyle l}\downarrow & & \downarrow{\scriptstyle i_K^H} \\ X(L) & \xrightarrow[i_L^H]{} & X(H) \end{array}$$

Adopt the temporary notational convention that if $x \in X(1)$, then x_L denotes the image of x under projection to $X(L)$. The pullback space Y consists of pairs $\{(x_L, x'_K) : x_H = x'_H\}$ and $Y = \coprod_h Y_h$, where h runs through double-coset representatives, and Y_h consists of pairs (x, xh) satisfying the preceding condition. The map sending $x_{L\cap K^h}$ to the pair $(x, xh) \in X_L \times X_K$ is then a homeomorphism of $X(L\cap K^h)$ onto Y_h, so that the preceding square can be identified with the following:

$$\begin{array}{ccc} \coprod_h X(L\cap K^h) & \xrightarrow[\coprod_h i_{L\cap K^h}^{K^h} c_h]{} & X(K) \\ {\scriptstyle \coprod_h i_{L\cap K^h}^{L}}\downarrow & & \downarrow{\scriptstyle i_K^H}. \\ X(L) & \xrightarrow[i_L^H]{} & X(H) \end{array}$$

Note: $i_{L\cap K^h}^{K^h} c_h = c_h i_{L^{h^{-1}}\cap K}^{K}$, and, as with ordinary cohomology, the preceding square is Cartesian because the decomposition $L = \bigcup_j \ell_{jh}(L\cap K^h)$ into $L\cap K^h$

cosets implies that:

$$Lh = \bigcup_j \ell_{jh}(Lh \cap hK),$$

$$LhK = \bigcup_j \ell_{jh}(LhK \cap hK) = \bigcup_j \ell_{jh} hK.$$

The union is disjoint and summing over h gives

$$H = \bigcup_{j,h} \ell_{jh} hK.$$

The remaining Mackey axioms, such as transitivity and compatibility of conjugation with i^* and i_* also follow easily from the definitions. Note: The cohomological triviality of induced inner automorphisms in $M4$ holds only for elements of H. ■

Next we cite an important formal consequence of the rules M and F. Let K_n be a set of subgroups of H. For $h \in H$, consider the maps:

$$i : X(K_n \cap K_m^h) \to X(K_n), \quad i = i_{K_n \cap K_m^h}^{K_n}$$

$$j : X(K_n \cap K_m^h) \to X(K_m), \quad j = i_{K_n^{h^{-1}} \cap K_m}^{K_m} c_h,$$

which combine to give a homomorphism:

$$'i_* - j'_* : \coprod_n h^*(X(K_n)) \to \coprod_{n,m,h} h^*(X(K_n \cap K_m^h)),$$

with h running over double-coset representatives for the pair H_n, H_m.

PROPOSITION 6.2. *Suppose that:*

$$\coprod_n i_* : \coprod_n \omega^*(X(K_n)) \to \omega^*(X(G))$$

is surjective, then:

$$0 \to h^*(X(G)) \to \coprod_n h^*(X(K_n)) \xrightarrow{'i_* - j'_*} \coprod_{m,n,g} h^*(X(K_n \cap K_m^g))$$

is an exact sequence.

Proof: Choose $\alpha_n \in \omega^0(X(K_n))$ with $\sum_n i_* \alpha_n = 1$. If $\{x_n\} \in \coprod_n h^*(X(K_n))$ belongs to the kernel of $'i_* - j'_*$, set $x = \sum_n i_*(\alpha_n x_n)$. Apply i_n^* to both sides, then apply M. The kernel assumption forces most terms to vanish; applying F we have $i_n^* x = \sum_m i_* \alpha_m x_n = x_n$ for each n. This proves exactness in the middle, and injectivity on the left is similar. ■

The assumption made in 5.4 holds if $\{K_n\}$ equals $\{G_p\}$, a set of representative p-Sylow subgroups of G. The degree then defines a homomorphism:

$$\deg : \omega^0(X) \longrightarrow \mathrm{Hom}(H_0(X), \mathbb{Z}).$$

For each covering map $i : X(K) \to X(G)$, $\deg(i_*(1)) = [G : K]$, the index, therefore:

$$i_* i^* : \omega^*(X(G)) \otimes \mathbb{Z}\left[\frac{1}{[G:K]}\right] \to \omega^*(X(G)) \otimes \mathbb{Z}\left[\frac{1}{[G:K]}\right]$$

is an isomorphism by Frobenius reciprocity. Hence we must have a surjection:

$$\coprod_p i_* : \coprod_p \omega^*(X(G_p)) \to \omega^*(X(G)).$$

In general if $g \in G$ normalizes the subgroup K, then c_g is an automorphism of $X(K)$. However even if g centralizes K, c_g need not be homotopic to the identity. This is the case if we take $X(G)$ to be a finite approximation to the classifying space BG, then pass to the limit. The universal cover $X(1)$ is contractible.

Proposition 6.3 is a very useful local form of this result.

Proposition 6.3. *Suppose that a p-Sylow subgroup G_p of G is abelian, then*

$$i^* : h^*(BG) \twoheadrightarrow h^*(BG_p)^{inv}$$

is surjective. Invariant means invariant with respect to the induced action of the quotient group $N_G(G_p)/C_G(G_p)$.

Proof: For any element $g \in G$, both G_p and G_p^g are contained in the centralizer C of $G_p \cap G_p^g$. Hence G_p and G_p^g are conjugate in C; i.e., there exists $t \in C$ with $tg \in N(G_p)$. Given the contractibility of universal covers:

$$c_{tg} i^* \simeq c_g i^*$$

the stability of an element can be described by the action of elements tg. ■

As a foretaste of a more general result, Corollary 6.1 is an easy consequence of the argument so far.

COROLLARY 6.1. *Let G be a split extension:*

$$1 \to C_m \to G \rightleftarrows C_n \to 1, \qquad (n,m) = 1,$$

such that the action of C_n on C_m is faithful, then:

$$h^*(BG) \cong h^*(BC_n) \oplus h^*(BC_m)^{\mathrm{inv}}$$

Again we do no more than generalize a familiar result from ordinary cohomology. We also have a variant of the theorem (Theorem 5.1) on detection by Chern classes.

THEOREM 6.1. *Let h^* be a complex-oriented cohomology theory as before, and let the p-Sylow subgroup G_p of G be cyclic. There exists a virtual representation ρ of G such that the restriction of the top-dimensional Chern class $c_d^h(\rho)$ generates the image of $h^*(BG)$ in $h^*(BG_p)$. Here $d = [N(G_p) : C(G_p)]$.*

Proof: Combine the representation theory used in proving the result for ordinary cohomology with the isomorphism:

$$h^*(BG_p) \cong h^*(\text{point})[[x]]/[p^t](x)$$

where the generator x is taken to be the first Chern class of a one-dimensional representation of G_p.

For an alternative proof using the notion of sparse coefficients, see [28]. ■

Propositions 6.2–6.3, Corollary 6.1, and Theorem 6.1 and the generalized version of Theorem 5.1 enable us to study the structure of $h^*(BG)$ in much the same way as ordinary cohomology. Corollary 6.1 is a special case of Theorem 6.2.

THEOREM 6.2. *Let G be a finite group such that G_p is cyclic for all primes p dividing $|G|$. Then if h^* is a complex-oriented cohomology theory, $h^*(BG)$ is generated by Chern classes.*

Proof: Restriction defines an injective map:

$$h^*(BG) \longrightarrow \coprod_{p||G|} h^*(BG_p),$$

and Theorem 6.1 applies. ■

EXAMPLE. At odd primes the assumptions of Theorem 6.1 are satisfied by the perfect group $SL_2(\mathbb{F}_p)$ of order $(p^2-1)p$. As in ordinary cohomology, if ρ_1 is the defining representation in characteristic $p, c_2^h(\rho_1)$ covers all primes dividing p^2-1. At the prime p we must use a cuspidal representation of degree $(p-1)/2$.

EXERCISE. Prove an analogous result for the sporadic simple group J_1 of order $8\cdot 3\cdot 5\cdot 7\cdot 11\cdot 19$.

The situation becomes more interesting when Sylow subgroups are no longer cyclic but still of restricted type. In what follows and with the Mathieu groups in mind, we assume that G_p is either noncyclic of order p^2, i.e., isomorphic to $C_p\times C_p$ or nonabelian of order p^3. In the first case an extension of the complex orientation argument for C_p yields Proposition 6.4.

PROPOSITION 6.4. *The ring*

$$h^*(B(C_p^A\times C_p^B)) = h^*\ \textit{(point)}\ [[x_1,x_2]]/([p](x_1),[p](x_2)),$$

and in even dimensions is generated by the first Chern classes of the pullbacks of representations $A\ (B)\mapsto e^{2\pi i/p}$, $B\ (A)\mapsto 1$.

Now suppose that $|G_p| = p^3$ and as usual exclude the case $p=2$. Let h^* be a quotient theory of Ω^*_{SO} (hence complex-oriented), which we localize at the prime p. Independently of whether G is metacyclic or elementary, we have a short exact sequence:

$$1\to C_p^C\to G\to C_p^A\times C_p^B\to 1.$$

For $G\cong(p_+^{1+2})$ the notation is taken from Chap. 5; for the metacyclic group C maps to the pth power of some preimage of A. The generalized spectral sequence of the extension, converging to h^* rather than to ordinary cohomology H^*, has E_2-page, given by:

$$E_2^{r,s} = H^r(B\langle A,B\rangle, h^s(B\langle C\rangle)),$$

so that summing over r,s we have $\mathbb{Z}/p[\alpha_1,\alpha_2]\otimes\wedge(\nu)\otimes h^*[[x]]/[p](x)$. Examining the spectral sequence allows us to prove an analog of Theorem 5.2.

Recall from Chaps. 2 and 3 that locally elliptic cohomology is closely related to a local version BP^* of complex cobordism. We now suppose that h^* has coefficients equal to $\mathbb{Z}_{(p)}[v_{j_1},v_{j_2},\ldots]$ with $j_1=n<j_2<j_3\cdots$. We then have Theorem 6.3.

THEOREM 6.3 (M. TEZUKA–N. YAGITA [110]).

$$h^*(BG) = h^*[\alpha_1,\alpha_2,\kappa_1,\ldots,\kappa_{p-1},\zeta]/\textit{(relations)}$$

where α_1, α_2 are the Chern classes of one-dimensional representations and the remaining generators are expressible in terms of the Chern classes of an irreducible p-dimensional representation. In particular $h^{odd}(BG) = 0$.

Note: In contrast to ordinary cohomology, the sequence of generators κ_j begins with index 1 rather than 2.

Sketch of proof: The spectral sequence of the extension behaves better than that for ordinary cohomology because $BP^*(BC_p)$ is even-dimensional and p-torsion free. From our earlier discussion in Chap. 5 the only nontrivial differentials are d_3 and d_{2p-1}. We observed already that the E_2-page is described by:

$$\mathbb{Z}/p[\alpha_1, \alpha_2] \times \wedge(\alpha) \times h^*[[x]]/[p](x),$$

where as before $[p](x)$ denotes adding p copies of x according to the formal group law for h^*. Degrees of α_1, α_2. and x equal 2; that of ν equals 3. Arguing as for H^*, we have

$$d_3 x = \widetilde{\nu} \text{ with } \nu = \lambda \widetilde{\nu} \operatorname{mod}(v_{j_1}, \ldots), \qquad \lambda \neq 0 \pmod{p}.$$

Furthermore $E_4^{*,*}$ is generated by $px, \ldots, px^{p-1}, \alpha_1, \alpha_2$, and $x^{p-1}\widetilde{\nu}$ over $h^*[[x^p]]/[p](x)$. Note: Terms survive on the 0- rather than 2-fiber line in degrees less than $2p$. That these terms are universal cycles follows by applying the double-coset rule. The normal subgroup generated by C is central, so that $i^* i_*$ gives multiplication by the scalar p. As before x^p corresponds to a pth Chern class. Schematically we have Fig. 6.1.

So far the argument is similar to Lewis's for ordinary cohomology, but we must pay attention to coefficients h^*, which are now graded. How do we know that the kernel of d_3 on the $2n$-fiber line is exhausted by combinations of monomials $\alpha_1^{j_1}\alpha_2^{j_2}x^{ps}$? Suppose that d eliminates an element z that is not of this type; then by subtracting a suitable sum of monomials we assume that:

$$z = (a_1 x^s + a_2 x^{s+1} + \cdots)\alpha_1^{j_1}\alpha_2^{j_2} + c\alpha_1^{j_1+1}\alpha_2^{j_2-1} + \cdots$$

with $a_1 \neq 0 \operatorname{mod} p$ in h^* and $s \neq 0 \pmod{p}$.

If a_1 belongs to the ideal generated by v_{j_1}, we have $s < p^{j_1}$, since the formal group law has:

$$[p](x) = v_{j_1} x^{p^{j_1}} + \text{higher terms},$$

mapping to zero in $E_3^{2n,*}$.

Differentiating we have

$$((sa_1 x^{s-1} + \cdots)\alpha_1^{j_1}\alpha_2^{j_2} + \cdots)\nu \neq 0,$$

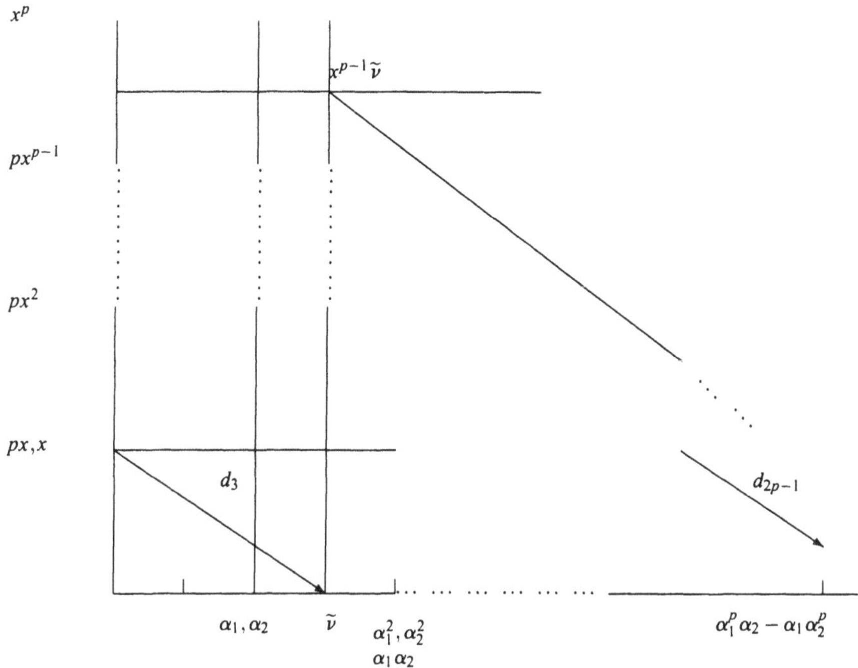

FIGURE 6.1. E_2-page of the spectral sequence (BP^*)

since sa_1 is still nonzero modulo p and $s-1 < p^{j_1}$. This contradicts our assumption that z lies in the kernel. The argument for the other fiber lines is similar (see [110, p. 399]).

As for H^*, we have now shown that in low degrees, $E_4 = E_\infty$. It remains to look at the differential d_{2p-1}. As 6.1 shows, this maps $x^{p-1}\tilde{\nu}$ to:

$$\alpha_1^p\alpha_2 - \alpha_1\alpha_2^p \in E_{2p-1}^{2p+2,0}.$$

This is the same argument as for ordinary cohomology [76, Lemma 6.20], and it depends on being able to describe this particular differential explicitly in terms of reduced power operations. Tezuka and Yagita[110] refer to this as an application of Kudo's transgression theorem.

Assuming this calculation we have now shown that all terms surviving to the E_{2p}-page have even degree, so the proof concludes as for H^*. Identifying the universal cycles px^j $(j = 1, \dots, p-1)$ with elements of the Chern subring is actually easier than before, since as we already noted, fiber multiplication by p is nonzero. The compatibility of induction with the evaluation of Newton polynomials in the classes c_k is also valid in the more general context of h^*. This concludes the sketch proof. ∎

The increased richness of the coefficients BP^* has the effect of complicating relations between generators just obtained. Particularly interesting are

$$\begin{aligned} \kappa_1 + v_1\zeta &= a_1 \\ p\kappa_2 + v_1\kappa_1\zeta &= a_2 \\ &\vdots \\ p\kappa_{p-1} + v_1 k_{p-2}\zeta &= a_{p-1}, \\ p^2\zeta + v_1\kappa_{p-1}\zeta &= a_p, \end{aligned}$$

where a_j belongs to the ideal generated by $(p^2, pv_1, v_1^2, v_2, \ldots, \alpha_1, \alpha_2)$. These relations are obtained [110, p. 401] by using the power series expansion for $[p](x)$ again.

For example $\kappa_1 + v_1\zeta$ is represented on the E_∞-page of the spectral sequence by $px + v_1x^p = [p](x)$ (modulo higher order terms). As with all spectral sequences, $\kappa_1 + v_1\zeta \mapsto px + v_1x^p$ corresponds to taking the quotient of two successive terms in the appropriate filtration of $BP^*(BG)$, that is to forgetting the higher order terms. The argument for other relations of this type is similar. Note: These explain both the disappearance of κ_1 from ordinary cohomology (heuristically map all v_js to 0) and the orders of $\kappa_2, \cdots, \kappa_{p-1}, \zeta$.

Combining everything done so far and specializing to the case of elliptic cohomology, localized at the odd primes $p = 3, 5, 7, 11, 23$, we obtain Theorem 6.4.

THEOREM 6.4. *$Ell^*(BM_{24})$ is concentrated in even dimensions.*

The same obviously holds for smaller Mathieu groups. Before discussing groups of order p^4, we must examine a wider class of groups and digress into yet another family of cohomology theories.

6.3. Morava K-Theories

The argument in Sec. 6.2 applied to both BP^* and to its structurally simpler quotients. Related to these are the so-called Morava K-theories (see [96, 97]). These have coefficients:

$\mathbb{F}_p[v_n, v_n^{-1}]$ for the theory $K(n)$ with dimension $v_n = 2(p^n - 1)$,
$\mathbb{F}[v_n]$ for the connective analog $k(n)$.

In the latter case we apply the Baas–Sullivan construction to the sequence $(p, v_1, v_2, \ldots, v_{n-1}, v_{n+1}, \ldots)$ as in Chap. 2, obtaining a family of classifying spaces or spectrum. By eliminating v_n as well, we obtain ordinary cohomology with $\mathbb{F}_p$-coefficients, and there are (families of) fibrations:

$$S^{2(p^n-1)}k(n) \underset{v_n}{\longrightarrow} k(n) \longrightarrow H(\ ,\mathbb{F}_p).$$

Iterating this map corresponds to inverting v_n in the coefficients, so we obtain the spectrum for the theory $K(n)$, formally

$$K(n) = \lim_{r\to\infty} S^{-2r(p^n-1)}k(n),$$

where $K(n)$ is periodic with period equal to $2(p^n-1)$. These theories are intermediate between $K(1)$, a summand of mod$p$ complex K-theory, and $H(\ ,\mathbb{F}_p)$; for example if $\dim(X) < 2p^n - 1$ and X is finite, $K(n)^*(X)$ is obtained from $H^*(X,\mathbb{F}_p)$ by tensoring over $\mathbb{F}_p$ with the coefficients. This follows, since the Atiyah–Hirzebruch spectral sequence has potentially nonzero differentials only in degrees $2r(p^n-1)+1$ (compare Sec. 3.2).

Among the other attractive properties of $K(n)$ are:

1. If $p \geqslant 3$, $K(n)^*(X)$ has a unique structure as a commutative graded ring.
2. $K(n)^* = \mathbb{F}_p[v_n, v_n^{-1}]$ is a graded field, and all graded modules over it are free.

   a. There are Künneth isomorphisms:

   $$K(n)^*(X\times Y) \cong K(n)^*(X) \underset{K(n)_*}{\otimes} K(n)^*(Y),$$

   b. There is a good Kronecker duality between homology and cohomology:

   $$K(n)^*(X) \cong \mathrm{Hom}_{K(n)_*}(K(n)_*(X), K(n)_*).$$

The formal group law gives rise to the simple and useful formula:

$$[p](x) = v_n x^{p^n}.$$

Hence it has height n, and at least over the algebraic closure $\overline{\mathbb{F}}_p$, it is isomorphic to a model height n formal group law, obtained by dividing in BP_* by the ideal generated by $(p, v_1, \ldots, v_{n-1})$, again inverting v_n. This at least suggests that the cohomology theory $B(n)^*$ obtained is such that $B(n)^*(X)$ is determined by $K(n)^*(X)$. Perhaps more significantly from our point of view, we can eliminate the sequence $(v_{n+1}, v_{n+2}, \ldots)$, and invert v_n to obtain a theory $E(n)^*$, satisfying $E(n)^*(X) = BP^*(X) \underset{BP_*}{\otimes} E(n)_*$. This is reminiscent of the situation described for elliptic cohomology in Chap. 2. We say more about this in Appendix A.

What does the general theory tell us about $K(n)^*(BG)$, with $|G| = p^n, n =$ small? If G is abelian the answer is given either by the complex orientation or the preceding Künneth formula. If G is nonabelian of order p^3, G is either metacyclic or elementary and the argument in Theorem 6.3 applies. Our approach was modeled on that of Lewis for p_+^{1+2}, but (modulo relations) applies equally well to the metacyclic case. Indeed in ordinary cohomology, the two cases are distinguished more by H^{odd}, with generators for a metacyclic group arising in dimension $2p+1$, for an elementary group in dimension 3. These do not occur in $K(n)^*$. As a special case of Theorem 6.3 we have Theorem 6.5.

THEOREM 6.5. *With coefficients extended from $\mathbb{F}_p$ to $\mathbb{F}_p[v_n, v_n^{-1}]$, $K(n)^*(BG)$ is generated by $\alpha_1, \alpha_2, \kappa_2, \ldots, \kappa_{p-1}, \zeta$. In particular $K(n)^{\text{odd}}(BG) = 0$.*

Note: κ_1 disappears as for ordinary cohomology, as a consequence of the relations. Those common to the two cases are

$$\alpha_1^p \alpha_2 = \alpha_1 \alpha_2^p, \alpha_1^{p^n} = \alpha_2^{p^n} = 0.$$

Remaining relations are obtained by specializing $\kappa_1 + v_1\zeta = a_1, \ldots$ from Theorem 6.3 and determining $a_1 \cdots a_p$ in the elementary and metacyclic cases. In p_+^{1+2} for example each $a_i = 0$, in accordance with the result obtained by mapping all v_j to 0, that is by passing from BP^* to ordinary cohomology with coefficients in $\mathbb{Z}_{(p)}$. However it is clear that the precise values do not affect our central claim that $K(n)^{\text{odd}}(BG) = 0$.

Groups of order p^3 are examples of groups G for which $K(n)^*(BG) = Ch_{K(n)}(G)$ and $K(n)^{\text{odd}}(BG) = 0$. In Sec. 6.4 we introduce a class of groups for which $K(n)^*(BG)$ is no longer generated by Chern classes but still satisfies the requirement that $K(n)^{\text{odd}}(BG)$ vanish.

DEFINITION ([57]). Let $Tre_{K(n)}(G)$ equal the subring of $K(n)^*(BG)$ generated by induced top-dimensional Chern classes of complex representations of subgroups $H \subseteq G$. The finite group G is *good* if:

$$K(n)^*(BG) = Tre_{K(n)}(G).$$

REMARK. The top-dimensional Chern class equals the Euler class of the representation. The notation *Tre* refers to Transferred Euler Classes.

6.4. Rank Two p-Groups

Recall that the *rank* of a p-group G equals the maximum $\mathbb{F}_p$-dimension of an elementary abelian subgroup of G. For $p \geqslant 3$ the group G has rank 1 if and only if it is cyclic. For $p \geqslant 5$ the group G has rank 2 if and only if it falls into one of three classes:

1. Metacyclic (not necessarily split).
2. Groups $G = C(r+2)$ with the same generators and relations as for (p_+^{1+2}) except that:

$$C^{p^r} = 1 \qquad [A,B] = C^{p^{r-1}},$$

giving $|G| = p^{r+2}$.

3. $G = G(r+3,e)$ with presentation:

$$\langle A,B,C| \quad A^p = B^p = C^{p^{r+1}} = 1, \quad [B,C] = 1, \quad [A,B^{-1}] = C^{p^r e}, \quad [A,C] = B \\ \text{with } e \neq 0 \bmod p\rangle.$$

Note: $G(r+3,e) \supseteq C(r+2)$ is a normal subgroup of index p. For details of this classification, see [22].

Arguing very much as in Sec. 6.3 we prove Theorem 6.6.

THEOREM 6.6. *Let $p \geqslant 5$ and let G be a p-group belonging to class 1 or 2. Then $BP^*(BG) = Ch_{BP}(G)$, and in particular $BP^{odd}(BG) = 0$.*

Proof: Consider first class 2. The spectral sequence of the central extension of $\langle C\rangle$ by $\langle A,B\rangle$ behaves very much as before, with the proviso that if $r \geqslant 2$, we must introduce a new two-dimensional generator $\gamma = \alpha_3$. This arises since the abelianized group of order p^3 has an extra generator corresponding to the class of C. For the argument in ordinary cohomology, see [113].

If G is metacyclic, the argument is easier, since the spectral sequence of the defining extension is trivial ($E_2^{\text{odd},*} = 0$) because the formal group law implies that $BP^*(BC_{p^r})$ is p-torsion free. To obtain precise generators, we need only look at invariant elements on the 0-fiber line, all of which can be expressed as Chern classes of induced representations. Clearly $BP^{odd}(BG) = 0$. ∎

In both cases it is actually not difficult to see that $BP^*(BG)$ equals $Tre_{BP}(G)$. We must reduce the dimension of the induced representation by varying the index of the supporting subgroup and by combining it with one-dimensional representations. This is essential for groups of class 3, but as a warm-up exercise, let us consider a metacyclic group of order p^3.

EXAMPLE/EXERCISE. Let the representation ξ be induced up from the normal subgroup $\langle A\rangle$ of index p. (We assume that generators A and B have orders p^2 and p, respectively, and $BAB^{-1} = A^{1+p}$). The Chern classes $c_j(\xi), 1 \leqslant j \leqslant p$ and β, equal to the first Chern class of the representation $\hat{\beta}(A \mapsto 1, B \mapsto e^{2\pi i/p})$ provide our initial BP^*-generators for $BP^*(BG)$. Next we replace $\langle A\rangle$ by $\langle A^p, B\rangle$, which is also maximal abelian and let η_s be induced up to the subgroup $\langle A^{p^{1-s}}, B\rangle$ from the representation $(A^p \mapsto e^{2\pi i/p}, B \mapsto 1)$, Then:

$$c_{p^s}(\eta_s) = x^{p^s} \pmod{(v_1, \ldots)}.$$

By the double coset formula applied to induction up to G followed by restriction down to $\langle A^p\rangle$, we have

$$i^* i_*(c_{p^s}(\eta_s)) = p^{1-s} x^{p^s} \pmod{(v_1, \ldots)}.$$

If we now write e for the top-dimensional class, the relation:

$$i_*(e(l\eta_s \oplus j\eta_0 \oplus k\hat{\beta})) = i_*(c_{p^{ls}}(\eta_s)^l) c_p(\eta_0)^j \beta^k$$

allows us to write down a new family of BP^*-module generators [122]. ■

We now turn to groups of Class 3, which can be described by means of an extension:

$$1 \to \langle A, B, C^p\rangle \to G \to \langle C\rangle \to 1.$$

For BP^* [or a suitable related theory, such as $K(n)^*$], we have the spectral sequence:

$$E_2^{*,*} = H^*(BC_p),\ BP^*(BC(r+2)) \Rightarrow BP^*(BG).$$

On the generators we have already found for $BP^*(BC(r+2))$, the quotient group acts by:

$$C^*\kappa_i = \kappa_i, \qquad C^*\alpha_1 = \alpha_1, \qquad C^*\alpha_2 = \alpha_2 -_{BP} \alpha_1,$$

where the suffix BP indicates that subtraction is according to the formal BP-group law. This action is clear from the presentation for G.

Write

$$\omega = \prod_{\lambda \in \mathbb{F}_p} (\alpha_2 +_{BP} [\lambda]\alpha_1) = \alpha_2^p + \cdots,$$

where the elements $\alpha_2^j \omega (j \geqslant 0)$ are invariant under C^*. Note the relation $\alpha_1 \omega = 0$ in $BP^*(BC(r+2))$ from earlier calculations for the smaller group.

THEOREM 6.7 ([122]). $BP^*(BC(r+3,e)) = Tre_{BP}(C(r+3,e))$.

Proof: This involves several steps.

STEP 1: Invariants $BP^*(BC(r+2))^{\langle C\rangle}$ are generated multiplicatively by:

$$1, \alpha_1, \kappa_1 \cdots \kappa_{p-1}, \zeta, \alpha_2^j \omega \quad (1 \leqslant j \leqslant p-1).$$

By subtracting elements that are easily seen to be invariant, we are reduced to considering the case:

$$z = \alpha_2^r \alpha_1^s \zeta^t + \cdots \qquad 1 \leqslant r \leqslant p-1.$$

Then:

$$\begin{aligned}(1-C^*)x &= ((\alpha_2 +_{BP} \alpha_1)^r - \alpha_2^r)\alpha_1^s \zeta^t + \cdots \\ &= r\alpha_2^{r-1}\alpha_1^{s+1}\zeta^t + \cdots \\ &\neq 0\end{aligned}$$

Hence z is not invariant.

Write $N = 1 + C^* + \cdots + C^{*p-1}$.

By the well-known calculation of the cohomology of a cyclic group:

$$E_2^{2n-1,*} = \ker N / \operatorname{im}(1-C^*).$$

STEP 2: $\ker N / \operatorname{im}(1-C^*) = 0$.

If the restriction i^*z to the subgroup $\langle B, C^p\rangle$ is nonzero, then $i^*(Nz) = pi^*(z) \neq 0$, since $BP^*(B\langle B, C^p\rangle)$ has no p-torsion; therefore $z \notin \ker N$.

Now let $i^*z = 0$.

From the fact that the spectral sequence for $C(r+2)$ collapses and relations in $BP^*(BC(r+2))$, z is a linear combination of the kind considered in Step 1, with $0 \leqslant r \leqslant p-1$, $s \geqslant 1$. By the previous calculation, if $r \neq p-1$:

$$\alpha_2^r \alpha_1^s \zeta^t + \cdots \in \operatorname{im}(1-C^*).$$

Then by the usual subtraction, we assume that $r = p-1$. For such a z:

$$\begin{aligned}Nz &= \sum_{\lambda \in \mathbb{F}_p} (\alpha_2 +_{BP} [\lambda]\alpha_1)^{p-1}\alpha_1^s \zeta^t \\ &= -\alpha_1^{s+p-1}\zeta^t + \cdots \\ &\neq 0 \ (\mathrm{mod}\ (p, v_1, \ldots)).\end{aligned}$$

STEP 3: All elements in $BP^*(BC(r+1))^{\langle C\rangle} = E_2^{0,*}$ are permanent cycles, $E_\infty = E_2$, and $BP^*(BG) = BP^{\mathrm{even}}(BG)$.

A corollary of the argument so far is that if we replace the theory BP^* by the theory with coefficients in $\mathbb{Z}_{(p)}[v_n, v_{n+1}, \ldots]$ (so that with further reduction mod$p$ we have $B(n)^*$ up to inversion of v_n), we can also prove that the connective Morava K-theories $k(n)^*$ follow the same pattern. This is enough to show that $K(n)^{\mathrm{odd}}(BG(r+3,e)) = 0$.

Combining what we proved so far, we know that $BP^*(BG)$ is BP^*-generated by:

$$1, \alpha_1, \alpha_2^j\omega, \kappa_1, \ldots, \kappa_{p-1}, \zeta, \eta,$$

where η corresponds to a nonzero element in $E_2^{2,0}$. We already showed that most of these elements belong to the Chern subring. Recalling that the letter x stands for the basic BP-Chern class for a cyclic group, we have Step 4.

STEP 4: If i denotes the inclusion of $\langle B, C\rangle$ in G, then:

$$i_*(x^{p-1}\alpha_2^j) \text{ restricts to } -\alpha_2^{p-1+j} \text{ on } \langle B\rangle \bmod (v_1, v_2 \cdots).$$

This is a messy but straightforward application of the double-coset formula (exercise).

STEP 5: $Tre_{BP}(G) = BP^*(BG)$.

If ρ is the representation of the subgroup $\langle B, C\rangle$ of index p, whose image under induction up to G gives the classes $\kappa_1, \ldots, \kappa_{p-1}, \zeta$, and $\widetilde{\beta}$ is the obvious one-dimensional representation, then $\alpha_2^j\omega$ is represented by:

$$i_*(e(p-1)\rho + j\widetilde{\beta})$$

and κ_j (modulo classes of lower degree) by:

$$i_*(e(j\rho)).$$

The letter e (Euler) again denotes the top-dimensional Chern class. The proof of Theorem 6.7 is now complete. ■

REMARK. The preceding argument also applies to ordinary cohomology, and it shows that $Tre_H G = H^{\mathrm{even}}(BG, \mathbb{Z})$. But it does not follow that $Ch(G) = Tre_H G$; indeed the classes $i_*(e(p-1)\rho + j\widetilde{\beta}) \notin Ch(G)$ for $z \leqslant j \leqslant p-2$; see [122, Lemma 4.16].

In essence this is a consequence of the failure of Chern classes to commute strictly with induction. The same elements also show that Grothendieck and skeletal filtrations on the representation ring $R(C(r+2,e))$ do not coincide for $p \geqslant 5$, see [122, Theorem 4.18].

Although this subsection is devoted to BP^* and its well-behaved related theories such as $K(n)^*$, we consider our main subject, which is elliptic cohomology. Using results on the elliptic genus, which are equivalent to Landweber exactness, in Chap. 2 we proved, that taking the tensor product over $\Omega^*_{SO} \otimes \mathbb{Z}[1/2]$ of cobordism with $\mathbb{Z}[1/2][\delta, \varepsilon, \varepsilon^{-1}]$ gives $Ell^*(X)$. Assuming as always that cobordism at the prime p can be described in terms of BP-theory, and applying Theorems 6.3, 6.6, and 6.7 for this theory, we have Complement 6.1.

COMPLEMENT 6.1. *Let $p \geqslant 5$ and G a p-group of rank equal to 2. Then $Ell^{odd}(BG) = 0$.*

The trick used in proving Theorem 6.7 for BP^* (and $K(n)^*$) is to present $G(r+3,e)$ as an extension:

$$1 \to C(r+2) \to G(r+3,e) \to C_p \to 1,$$

and use the collapsing of the associated spectral sequence, and the goodness of $C(r+2)$ to prove the goodness of $G(r+3,e)$. This technique can be generalized, as we indicate briefly.

At the end of our introduction to the Morava K-theories, we introduced the theories $B(n)^*$. Related to these we have $\widetilde{P}(n)^*$ with coefficients in $\mathbb{Z}_{(p)}[v_n, v_{n+1}, \ldots]$ and $P(n)^*$ obtained by further reduction modulo p. Let the p-group G be presented as an extension:

$$1 \to G_1 \to G \to C_{p^s} \to 1.$$

Suppose that for all $n \geqslant 0$, G_1 satisfies the conditions:

$$\left.\begin{array}{ll} 1. & P(n)^*(BG_1) \cong BP^*(BG_1) \underset{BP^*}{\otimes} P(n)^*. \\ 2. & \text{The } \widetilde{P}(n)\text{-spectral sequence of the extension} \\ & \text{is such that } E_2^{\text{odd},*} = 0. \end{array}\right\} \qquad (\sharp)$$

We then have Theorem 6.8.

THEOREM 6.8. *If the normal subgroup G_1 satisfies both conditions in ($\sharp$) then G satisfies Condition 1 for all $n \geqslant 0$.*

Proof: See [122]. ∎

As a consequence of this theorem, we have the Künneth formula:

$$P(n)^*(B(G \times H)) = P(n)^*(BG) \underset{P(n)^*}{\otimes} P(n)^*(BH)$$

for all finite p-groups H, so we can show that:

$$K(n)^*(BG) \cong BP^*(BG) \underset{BP^*}{\otimes} K(n)^*.$$

It follows that all p-groups of rank equal to 2 ($p \geqslant 5$) are good.

In the following subsection, we need a very special case of Theorem 6.8, which we state separately as Proposition 6.5.

PROPOSITION 6.5. *Let G be an extension of the elementary abelian group G_1 by the cyclic group C_p. Then $BP^*(BG)$ is generated in even dimensions.*

Sketch of proof: Since $BP^*(BC_p)$ is p-torsion free and the Künneth formula holds for abelian groups, $BP^*(BG_1)$ is p-torsion free, hence:

$$E_2^{\mathrm{odd},*} = H^{\mathrm{odd}}(C_p, BP^*(BG_1)) = 0,$$

and the spectral sequence collapses. As in Theorems 6.3, 6.5–6.7 we can use $E_2^{2,0}$ and the invariant elements in $E^{0,*}$ to construct a family of even-dimensional generators for $BP^*(BG)$. ■

6.5. Groups of Order p^4

If p is an odd prime, the nonabelian groups of order p^4 can be listed as follows in terms of at most four generators A, B, C, D satisfying the given relations:

1. $A^{p^3} = B^p = 1, A^B = A^{1+p^2}$, metacyclic.

2. $C^{p^2} = B^p = A^p = [C,B] = [A,C] = 1, [B,A] = C^p$.

3. $A^{p^2} = B^{p^2} = 1, A^B = A^{1+p}$, metacyclic.

4. (metacyclic of order p^3) $\times$ (cyclic of order p).

5. $A^{p^2} = B^p = C^p = [A,B] = [B,C] = 1, [A,C] = B$.

6–8. $A^{p^2} = B^p = [B,C] = 1, [A,C] = B, A^B = A^{1+p}, C^p = A^{sp}$, where $s = 0, 1$ or a quadratic nonresidue modulo p.

9. (elementary nonabelian of order p^3) $\times$ (cyclic of order p).

10. For $p \geqslant 5$, this group has four generators and exponent p:

$$[A,B] = [A,C] = [A,D] = [B,C] = 1, \qquad [C,D] = B, \qquad [B,D] = A.$$

For $p = 3$, we obtain the Wreath product $\mathbb{Z}/3 \wr \mathbb{Z}/3$.

This list is taken from [30, pp. 87–88]. Another less explicit reference is [59, Chap. II, Satz 12.6], where the classification is obtained in terms of Blackburn's theorem on p-groups of rank 2, already quoted. This covers Cases 1, 2, 3, 7, and 8, and Cases 4 and 9 are direct products. The degenerate Case 6 (with $s = 0$) is an extension of the elementary abelian group generated by A^p, B, and C by a cyclic group of order p. The same holds for the minimal nonabelian group Case 5, and the last case (10) contains the normal elementary abelian subgroup $\langle A,B,C\rangle$, as does the Wreath product in the special case $p = 3$.

It should now be clear that we have proved Theorem 6.9.

THEOREM 6.9. *Let p be an odd prime and G a group of order p^4. Then $BP^*(BG)$ is generated in even dimensions.*

Using the work of Tezuka and Yagita [110, 111] outlined in Sec. 6.3 the same holds for the Morava K-theories $K(n)^*$. Before the reader becomes too ambitious we note the following counterexample.

COUNTEREXAMPLE ([68]). Let G be a 3-Sylow subgroup of $Gl_4(\mathbb{F}_3)$, then $K(2)^{\text{odd}}BG \neq 0$.

The argument has since been extended to all odd primes p by a student, K. Lee of I. Kriz. These counterexamples are of order p^6, although the method and the classification of groups of order p^5 suggest that the exponent can be reduced by 1.

6.6. Notes

The concept of a module over a ring-valued functor satisfying Frobenius reciprocity and the double-coset rule goes back to the early days of algebraic K-theory, and it has proved its worth in numerous applications, such as surgery theory. For a systematic development, including more detail on the Burnside ring, see [40], and also the promised book (containing many examples) by J. Greenlees. As the text makes clear, provided certain axioms are satisfied, a representation theoretic object for the group G is detected on a suitable family of subgroups $\{H_i : i \in I\}$ contained in G. If we do not mind introducing denominators, abelian subgroups H suffice; for more delicate results we need elementary or hyperelementary subgroups. This goes some way to explain the importance of the Sylow subgroup structure of G.

The other major piece of theory on which the results of Chap. 6 rest is Brown–Peterson cohomology, which we have rather cavalierly linked with cobordism localized at a prime p. Formally a homotopy theorist says that the local spectrum $MU_{(p)}$ splits as the sum of suspensions of BP:

$$MU_{(p)} \simeq \bigvee_i S^{n(i)}BP,$$

with

$$\lim_{i\to\infty} n(i) = \infty.$$

We give the main definitions and results that we used in Appendix A. Our account at least explains how generators v_i actually arise in degree $2(p^i - i)$. But the reader is strongly advised to consult [1] for an extended and elegant treatment.

In Chap. 7 we shall see that for good finite groups G, the elliptic cohomology of BG is given by completing an elliptic representation ring. This holds for example for groups of odd order where $|G_p| = p^t\ (t \leqslant 4)$ for all primes $p \,|\, |G|$, and it hinges on the fact that $Ell^{\mathrm{odd}}(BG_p) = 0$.

Now consider our permanent test case of the Mathieu group M_{24}; at odd primes the Sylow structure is good, but the order is even. Heuristically we can say two things:

1. The completion result holds for the level-one theory with coefficients in $\mathbb{Z}[1/6][g_2, g_3]$, where g_2 and g_3 (interpreted as modular forms) are the familiar functions from the theory of elliptic curves. Inverting both 2 and 3 in the coefficients brings Theorem 6.1 into play, and for level 1 $Ell^*(BM_{24})$ is generated as in ordinary cohomology by the Chern classes of the Todd representation.

2. For level 2 the picture is not so clear. D. Green's argument for $H^*(BM_{24}, \mathbb{Z})_{\mathrm{odd}}$ (Theorem 5.3) cannot be carried over directly to BP^*, partly because of the more general coefficients, more seriously because of the class κ_1, which as we saw, maps to zero in H^*. In the case of elliptic cohomology, the Todd representation certainly plays an important role: Only the prime 3 matters because of Comment 1. The Tezuka–Yagita calculations for $K(2)^*$ show that κ_1 disappears, so that up to checking the behavior of graded field coefficients, Green's argument does apply. $K(1)^*$ is a summand of K-theory, and it equals its Chern ring; that is, $K(1)^*(BM_{24})$ is generated by the classes of the Todd representation at odd primes.

I. Kriz's counterexample to the conjecture that $K(n)^{\mathrm{odd}}(BG) = 0$ for all finite groups G can be extended from $p = 3$ to all odd primes [75]. The groups concerned $C_p^4 \rtimes C_p^2$ are also of interest because these are closely related to the p-Sylow subgroups of Chevalley groups of composite order, for which the conjecture holds. Thus [109] has proved Theorem 6.10.

THEOREM 6.10. *Let $G(\mathbb{F}_q)$ be the group of $\mathbb{F}_q (q = l^t)$ points of a connected reductive $\mathbb{Z}$-group scheme, such that $H^*(G(\mathbb{C}), \mathbb{Z})$ has no p-torsion. If $K(n)^*$ denotes Morava K-theory for the prime $p(\neq l)$, then:*

$$K(n)^{odd}(BG(\mathbb{F}_q)) = 0.$$

For an example of this result, consider the special case of $E_8(\mathbb{F}_3)$, which contains the Thompson group of order equal to:

$$2^{15}\, 3^{10}\, 5^3\, 7^2\, 13\, 19\, 31$$

and take $p = 13, l = q = 3$ in Theorem 6.10. Tanabe's argument shows that $K(n)^*(BE_8)$ is actually generated by Chern classes of degrees 12, 18, 24, and 30. Furthermore $c_{12}(248)$ restricts to the generator of $K(n)^*(BTh)$, providing yet another illustration of Theorems 5.1 and 6.1. (See [35, pp. 176–177] for more numerical information.)

Kriz–Lee results show that the bad prime for the Morava K-theory equals the natural characteristic for the group G. Their examples involve the construction of nonzero elements in:

$$H^1(C_p, \widetilde{K}(n)^{\text{even}}(BG_1)),$$

where G_1 is a good normal subgroup of G and $\widetilde{K}(n)$ is an integral version of $K(n)$. This means that we replace $\mathbb{F}_p$ by integers in a degree n unramified extension of $\mathbb{Q}_p$. The discovery of such elements is clearly of great importance.

7

Completion Theorems

Chapter 7 gives an algebraic description of:

$$Ell^*(EG \underset{G}{\times} X) \otimes \mathbb{Z}\left[\frac{1}{|G|}\right],$$

when X is a finite CW-complex admitting an action by the finite group G. We are particularly interested in the case when X is a point and $EG \underset{G}{\times} X = BG$. The completion theorem for K-theory, mentioned in Chap. 3, suggests that $Ell^*(BG) \times \mathbb{Z}[1/|G|]$ is isomorphic to a suitable completion of an evenly graded elliptic character ring $\mathcal{E}ll^*_G$. In general this is the best for which we can hope, since we showed that $Ell^{\text{odd}}(BG)$ need not vanish, as a consequence of the nontriviality of the spectral sequence of an extension:

$$1 \to G_1 \to G \to C_p \to 1.$$

Since triviality can be reintroduced at the price of tensoring with $\mathbb{Z}[1/|G|]$, Atiyah's original proof for solvable groups G can be saved. To avoid introducing denominators in this way, we restrict the class of groups considered. Calculations in Chap. 6 show that interesting cases do exist, for example p-groups of rank equal to 2.

To prove a completion theorem for arbitrary groups, we introduce an equivariant elliptic cohomology theory. For this we follow [39, 38], first defining coefficients $\mathcal{E}ll^*_G$, then a structural map:

$$\Phi_G : MSO^*_G \to \mathcal{E}ll^*_G,$$

generalizing the elliptic genus, where MSO^*_G denotes an oriented equivariant cobordism ring.

Definition. $\mathcal{E}ll^*_G(X) = MSO^*_G(X) \underset{\Phi_G}{\otimes} \mathcal{E}ll^*_G$.

Here X is a finite G–CW-complex, and the script notation is chosen to emphasize that we introduced an approximation to $Ell_G^*(X)$ with denominators. However for this new theory, it is possible to use standard equivariant methods to prove that:

$$\mathcal{E}ll_G^*(X)_I^\wedge \cong \mathcal{E}ll_G^*(EG \times X).$$

At the present state of our knowledge, this result generalizes the K_G-completion theorem. Completion is with respect to powers of the ideal:

$$I = \ker\{i^* : \mathcal{E}ll_G^* \to \mathcal{E}ll_1^* = Ell^* .\}$$

Throughout Chap. 7, we assume that G has odd order.

7.1. Equivariant Coefficient Ring $\mathcal{E}ll_G^*$

Recall some notation from Chap. 2; $H = H_1$ denotes the complex upper half-plane and:

$$\Gamma_0(2) = \left\{ A = \begin{pmatrix} a & b \\ c & d \end{pmatrix} \in SL_2(\mathbb{Z}) : c \equiv 0 \pmod 2 \right\}.$$

As in Chap. 3 we consider pairs of commuting elements in G to write

$$G^{(2)} = \{(g_1, g_2) \in G \times G : g_1 g_2 = g_2 g_1\}.$$

The group $\Gamma_0(2) \times G$ acts on the left of $G^{(2)} \times H$ by:

$$\left(\begin{pmatrix} a & b \\ c & d \end{pmatrix}, g \right) \times ((g_1, g_2), \tau) \underset{\rho}{\mapsto} \left(g(g_1^d g_2^{-c}, g_1^{-b} g_2^a) g^{-1}, \frac{a\tau + b}{c\tau + d} \right).$$

Note: $A^{-1} = \left(\begin{smallmatrix} d & -b \\ -c & a \end{smallmatrix} \right)$ and the action on the pair (g_1, g_2) is chosen to mimic $(g_1, g_2)A^{-1}$.

For each $k \in \mathbb{Z}$ there is an induced action on the ring of holomorphic functions $\vartheta : G^{(2)} \times H \to \mathbb{C}$ defined by:

$$\vartheta(\cdot,\cdot) \underset{\rho_k}{\mapsto} (c\tau + d)^{-k} \vartheta(\rho(\cdot,\cdot)).$$

Here holomorphic means that for each fixed pair (g_1, g_2), ϑ is holomorphic in the usual sense.

DEFINITION. The abelian group $\mathcal{E}ll_G^{-2k}$ consists of all holomorphic functions $\vartheta : G^{(2)} \times H \to \mathbb{C}$ that satisfy

Dev-1: ϑ is invariant under the induced action ρ_k.

Dev-2: For each fixed pair (g_1,g_2), functions ϑ and ϑ' where $(\vartheta'(\cdot,\tau) = \tau^{-k}\vartheta(\cdot,-1/\tau))$, have power series expansions at the cusp $i\infty$ of the form:

$$\vartheta(\cdot,\tau) = \sum_{n\geqslant K} a_n q^{n/|g_1|}, \qquad \vartheta'(\cdot,\tau) = \sum_{n\geqslant K} b_n q^{n/|g_1|},$$

with $k \in \mathbb{Z}$, $q = e^{2\pi i/\tau}$, and $a_n, b_n \in \mathbb{Z}\left[1/2, 1/|G|, e^{2\pi i/|g_1 g_2|}\right]$. Note: $\left(\begin{smallmatrix} 0 & -1 \\ 1 & 0 \end{smallmatrix}\right) \notin \Gamma_0(2)$ and maps $\tau = i\infty$ to $\tau = 0$; also Dev-2 is the strongest possible integrality condition that can be imposed from the point of view of modular forms. This follows from the theory of level structures and the Tate curve (see [64]).

Dev-3: Let $C_{g_1}(G)$ be the centralizer of g_1 and let $\zeta_1 = \exp\{2\pi i/|C_{g_1}(G)|\}$. If n and $|C_{g_1}(G)|$ are coprime and σ_n denotes the automorphism:

$$\zeta_1 \mapsto \zeta_1^n$$

of the ring $\mathbb{Z}[1/|G|, \zeta_1]$, then:

$$\sigma_n(a_m(g_1,g_2)) = a_m(g_1,g_2^n)$$
$$\sigma_n(b_m(g_1,g_2)) = b_m(g_1,g_2^n).$$

The group structure in $\mathcal{E}ll_G^*$ is induced by the sum of functions, and the graded product by their product.

The Galois invariance condition (Dev-3) implies that by fixing g_1, functions $a_m(g_1,-)$ and $b_m(g_1,-)$ become characters of the centralizer $C_{g_1}(G)$, with coefficients extended from $\mathbb{Z}$ to $\mathbb{Q}$ [57, Proposition 1.5]. Condition Dev-2 and properties of class functions show that we need invert only the order of G. Hence up to the introduction of a restricted class of denominators, Condition Dev-3 is equivalent to requiring Fourier expansions of modular forms at the two cusps to be graded characters of $C_{g_1}(G)$.

Based on discussions in Chaps. 3 and 5, the family of rings $\{\mathcal{E}ll^*(H) : H \subseteq G\}$ describes the values of a Green functor. All we have to do (compare Sec. 3.1) is provide an induction formula for a pair of subgroups $H \subseteq K$:

$$i_*\vartheta\left((g_1,g_2),\tau\right) = \sum_{gH} \vartheta\left((g_1^g,g_2^g),\tau\right)$$

where $(g_1,g_2) \in K^{(2)}$ and the sum is taken over cosets gH in K such that $g_j gH \subseteq gH$ for $j = 1,2$.

The fact that i_* is well-defined follows from the invariance of θ under conjugation by an element k belonging to the smaller subgroup H. The definition in terms of sums and conjugation shows that Devoto axioms hold for $i_*\vartheta$, and Mackey and Green axioms can be checked as for ordinary representations. As an example we have Lemma 7.1.

LEMMA 7.1. *Let K,H be two subgroups of G, then induction from H followed by restriction to K satisfies the relation:*

$$i^*i_*\vartheta(g_1,g_2,\tau) = \sum_{g\in H\backslash G/K} \operatorname{ind}^K_{K\cap H^g} \operatorname{res}^{H^g}_{K\cap H^g}(\vartheta c_g)(g_1,g_2,\tau).$$

Proof: This follows the lines of the double-coset argument in Chap. 6 (Proposition 6.1) and depends on the (1–1) correspondence between H-cosets in G and the disjoint union:

$$\coprod_{g\in H\backslash G/K} K\cap H^g\backslash K.$$

As a Green functor $\mathcal{E}ll^*_G$ is a module over the Burnside ring $A(G)$. We already mentioned the universal property of $A(G)$, but as in the earlier case of equivariant cohomology, we can make the multiplication $A(G)\times\mathcal{E}ll^*_G\to\mathcal{E}ll^*_G$ explicit.

It is enough to consider finite G-sets of the form G/H, when:

$$[G/H]\vartheta(g_1,g_2,\tau) = \sum_{Hg}\vartheta(g_1^g,g_2^g,\tau) = \#(H,g_1,g_2)\vartheta(g_1,g_2,\tau).$$

Here the sum is taken over cosets belonging to the subset of $H\backslash G$ invariant under g_1g_2, and $\#(H,g_1,g_2)$ denotes their number. ■

In our first discussion of $A(G)$, we mentioned the decomposition of the unit element 1 as a sum of idempotents e_H, one for each conjugacy class of subgroups (H) in G. It is possible to prove (compare the use of Quillen descent in Chap. 3 and [39, Lemma 3.10]) that the important subgroups are rank 2 abelian.

LEMMA 7.2. *With respect to the $A(G)$-module structure on $\mathcal{E}ll^*_G$, $e_H\mathcal{E}ll^*_G=0$ unless $H=\langle g_1,g_2\rangle$ for some pair of commuting elements $g_1,g_2\in G^{(2)}$.*

The proof depends on knowledge of the idempotents e_H, properties of the Möbius function for a partially ordered set, and the definition of the $A(G)$-action just cited. Accepting this, the Mackey and Green axioms easily imply

PROPOSITION 7.1. *Restriction to two-generator abelian subgroups $\langle g_1,g_2\rangle$ of G induces an isomorphism:*

$$\mathcal{E}ll^*_G \xrightarrow[i^*]{} \lim_{\longleftarrow \langle g_1,g_2\rangle} \mathcal{E}ll^*_{\langle g_1,g_2\rangle},$$

where the limit is taken over inclusions twisted by conjugations.

Note: We can reduce the category of all abelian subgroups $\mathfrak{A}(G)$ used in Chap. 3 because of Lemma 7.2.

COROLLARY 7.1.

1. *Induction induces an epimorphism:*

$$\varinjlim_{\langle g_1,g_2\rangle} \mathcal{E}ll^*_{\langle g_1,g_2\rangle} \twoheadrightarrow \mathcal{E}ll^*_G,$$

2. *Summing over conjugacy classes (H) of subgroups $H = \langle g_1,g_2\rangle$ gives an isomorphism:*

$$\mathcal{E}ll^*_G \cong \bigoplus_{(H)} (\mathcal{E}ll^*_H)^{W(H)},$$

where W(H) is the Weyl Group introduced in Chap. 3.

REMARK. All we have done here is to restate the detection theorems used by [56] in a slightly more general setting.

We now have parallel theories for $\{\mathcal{E}ll^*_H : H \subseteq G\}$ and $\{\mathcal{E}ll^*(BH) : H \subseteq G\}$ but as yet no connecting family of maps corresponding to the flat-bundle homomorphism of K-theory. Help in this direction is provided by an alternative description of the coefficients $\mathcal{E}ll^*_G$, when G is an abelian group of rank 1 or 2. For such a group G:

$$C_g(G) = G \text{ for all } g \in G, \qquad G^{(2)} = G \times G.$$

Let $i : (\mathbb{Z}/|G|)^* \to GL_2(\mathbb{Z}/|G|)$ be the homomorphism:

$$n \mapsto \begin{pmatrix} n & 0 \\ 0 & 1 \end{pmatrix}.$$

With this notation group actions in Devoto's axioms are induced by the $GL_2(\mathbb{Z}/|G|)$ action on $G(2)$, given by:

$$(A,(g_1,g_2)) \mapsto (g_1^d g_2^{-c}, g_1^{-b} g_2^a).$$

Let $[g_1,g_2]$ represent the orbit of (g_1,g_2) under this action, and write $\Gamma([g_1,g_2])$ for the stabilizer subgroup in $\Gamma_0(2)$.

DEFINITION. Let $\mathcal{E}ll^{-2k}(\Gamma[g_1,g_2])$ be the group of holomorphic functions $\vartheta : H_1 \to \mathbb{C}$ that satisfy

Mod-1: $\vartheta(\tau) = (c\tau + d)^{-k}\vartheta(A\tau)$ for all $A \in \Gamma([g_1,g_2])$.

Mod-2: If τ_0 is a cusp for the subgroup $\Gamma([g_1,g_2])$ with $A(i\infty)=\tau_0$, then $\vartheta(\tau)$ and $\vartheta'(\tau)=(c\tau+d)^{-k}\vartheta(A\tau)$ both have power series expansions of the form given in Dev-2.

We write

$$\mathcal{E}ll^*(\Gamma[g_1,g_2]) = \bigoplus_k \mathcal{E}ll^{-2k}(\Gamma[g_1,g_2])$$

with Galois action σ by GL_2; let

$$\Lambda : \mathcal{E}ll^*_G \to \bigoplus_{[g_1,g_2]} \mathcal{E}ll^*(\Gamma([g_1,g_2]))$$

be the sum of the natural ring projections fixing the first coordinate (g_1,g_2) in ϑ.

PROPOSITION 7.2. *The homomorphism Λ is an isomorphism.*

Proof: We define an inverse Λ^{-1}. If $(h_1,h_2)\in G^{(2)}$, we can find a pair $[g_1,g_2]$ and a matrix B such that $B[g_1,g_2]=(h_1,h_2)$. Furthermore we can find $A\in\Gamma_0(2)$, projecting to $\overline{A}$ in $GL_2(\mathbb{Z}/|G|)$, such that:

$$B = i(\det B)\overline{A}.$$

The natural definition is now:

$$(\Lambda^{-1}\vartheta)((h_1,h_2),\tau) = \sigma(i(\det B))((c\tau+d)^{-k}\vartheta_{[g_1,g_2]}(A\tau)).$$

$\Gamma([g_1,g_2])$ invariance implies that Λ^{-1} is well-defined, and the restriction on the Fourier coefficients in Mod-2 implies that these also hold for $\Lambda^{-1}\vartheta$ (Dev-2).

Let us take a closer look at the case $C_p=\langle 1,g\rangle$, with $g^p=1$. $\Gamma_0(2)$ has two orbits in $C_p^{(2)}$, represented by $(0,0)$ and $(1,0)$. The preceding description shows that:

$$\mathcal{E}ll^*_{C_p} \cong \mathcal{E}ll^* \bigoplus \mathcal{E}ll^*(\Gamma_1(p)\cap\Gamma_0(2)).$$

Here the stabilizer subgroup $\Gamma_1(p)$ of $(1,0)$ consists of matrices $\left(\begin{smallmatrix}1&0\\c&1\end{smallmatrix}\right)$ with $c\equiv 0 \pmod p$. ■

With Proposition 7.2 we return to the classical theory of modular forms, so we can express $\mathcal{E}ll^*_{C_p}$ as a quotient:

$$\mathbb{Z}\left[\frac{1}{2}\right][\delta,\varepsilon,\Delta^{-1}][x]/\langle T_p(x)\rangle.$$

The full theory for this is given in [61]; T_p is divisible by a polynomial $\Phi_p(x)$ of cyclotomic type. This polynomial allows us to map $\mathcal{E}ll^*_{C_p}$ to $Ell(BC_p)$ in the same way that Atiyah maps $R(C_p)$ to $K(BC_p)$. Completion with respect to the argumentation ideal I gives an isomorphism in both cases. Let us recall the K-theoretic argument.

Complex orientation implies for the symmetric lens space $L^{2n+1}(p;1,\ldots,1)$:

$$K(L^{2n+1}) \cong \mathbb{Z}[x]/\langle x^{n+1}, [p]_K(x)\rangle,$$

with $[p]_K(x) = (1+x)^p - 1$.

The last relation follows by iterating the formal group law $x +_K y = x+y+xy$ p times. The reduced group:

$$\widetilde{K}(L^{2n+1}) \cong (\mathbb{Z}/p^{s+1})^r + (\mathbb{Z}/p^s)^{p-r-1},$$

with

$$n = s(p-1)+r \quad (0 \leqslant r < p-1),$$

contains p^n elements, and it is generated by $x, x^2, \ldots, x^{p-1}$, corresponding to the irreducible nontrivial representations of the cyclic group C_p. Taking the limit as n tends to ∞ shows that $K(BC_p)$ is isomorphic to $R(C_p)^\wedge_p$, the p-adic completion. The extra summand comes from the trivial representation. Topologically p-adic and I-adic filtrations coincide. To see this, let ρ be the regular representation and $\xi \in I$. Then $\rho\xi = 0$, since the character of ρ vanishes except at the identity, and the character of ξ vanishes at the identity. Thus $(p^n - \rho)\xi = p^n\xi$ and $p^n I \subseteq I^2$.

The other way round, we use the Adams operation ψ^{p^n}. We have $0 = \psi^{p^n}(\xi) = \xi^{p^n} + p\eta$, so that $\xi^{p^n} = -p\eta$ and $I^{p^n} \subseteq pI$.

The critical step in this argument is the formula $[p]_K(x) = (1+x)^p - 1 = x\left(\sum_{j=0}^{p-1}(1+x)^j\right)$, which must be replaced by the corresponding formula in elliptic cohomology. This comes close to being made explicit in [61] on the transformation theory of elliptic functions. See also [38] for a too rapid account of elliptic curves over arbitrary schemes, which at least explains the relation between modular forms of level $\Gamma(p)$ and the structure sheaf of the universal curve representing the moduli problem. This gives $T_p(x)$, which by [61] satisfies relations:

$$[p]_{Ell}(x) = T_p(x)F_p^{-1}(x), \qquad T_p(x) = x^{p^2}F_p(x^{-1}).$$

The form of the polynomial $F_p(x)$ [61, p. 441] suggests that of the divisor $\Phi_p(x)$ of $T_p(x)$ corresponding to the pth cyclotomic polynomial:

$$1+y+\cdots+y^{p-1} = \prod_{j=1}^{p-1}(y-\zeta^j),$$

with ζ equal to a primitive root.

In the elliptic case, we let δ be the structural constant in the equation for the Jacobi quartic:

$$y^2 = 1 - 2\delta x^2 + x^4.$$

(If δ is given a specific numerical value, this should be transcendental over $\mathbb{Q}$.)

If a denotes a point of the curve of order p, write

$$\Phi_p(x) = \prod_{a=\text{primitive}} (x - x(a)),$$

where $x(a)$ denotes the x-coordinate of a. This formula generalizes that of the cyclotomic formula relevant to the preceding K-theory argument, but it is more difficult to write explicitly, however $\deg \Phi_p(x) = p^2 - 1$ and $\Phi_p(0) = \pm p$, depending on the congruence class of p modulo 4.

As in K-theory we must compare the quotient of the polynomial ring $R[x]/T_p(x)$ to the quotient of the power series ring $R[[x]]/[p](x)$, $R = \mathbb{Z}[1/2][\delta, \epsilon, \Delta^{-1}]$. This involves inverting $F_p(x)$, i.e., an I-adic completion process. Note: By Proposition 7.2 the kernel of the elliptic augmentation can be identified with a ring of higher level modular forms to which, by inspecting the zeros, as described by [61], $F_p(x)$ belongs. Thus we have Proposition 7.3.

PROPOSITION 7.3. *If* $\hat{\cdot}$, *denotes* I*-adic completion with respect to* $I = \ker(\mathcal{E}ll^*_{C_p} \to \mathcal{E}ll^*_1)$:

$$(\mathcal{E}ll^*_{C_p})^\wedge \cong Ell^*(BC_p).$$

REMARK. Compare adjoining roots of $[p](x) = 0$ to R in the argument in Sec. 3.4.

How does Proposition 7.3 generalize? For elementary abelian groups, we can combine 7.2 and the isomorphism (Sec. 3.4):

$$Ell(B(C_p \times \cdots \times C_p)) \cong R[[x_1, \ldots, x_n]]/\langle [p](x_1) \cdots [p](x_n) \rangle.$$

For groups of order p^3 and p^4, we use methods in Chap. 6, and restrict both $\mathcal{E}ll^*_G$ and $Ell^*(BG)$ to subgroups of rank 2. This works because we already showed that if G is p-group of order p^t with $t \leqslant 4$ ($p =$ odd), then $Ell^{\text{odd}}(BG) = 0$. And without some weakening assumption, such as tensoring with $\mathbb{Z}(1/|G|)$, this argument fails in general.

Note in passing that calculations for a nonabelian elementary group of order p^3 in Chaps. 5 and 6 can be made in terms of the spectral sequence of the extension of $\langle B, C\rangle$ by $\langle A\rangle$ (or equally well of $\langle A, C\rangle$ by $\langle B\rangle$). In the case of Brown–Peterson theory, due to of the absence of p-torsion from coefficients $BP^*(B(C_p \times C_p))$, the structure of the spectral sequence is even simpler than for ordinary cohomology [75, Sec. 6.4].

In essence we use the first method of [13] to prove the K-theoretic completion theorem, in so far as this applies to solvable groups. In our case this is not a restriction, since $|G| = $ odd and the Feit–Thompson theorem applies (see [38]).

THEOREM 7.1. *If G is a finite group of odd order, then:*

$$Ell^*(BG)\otimes\mathbb{Z}\left[\frac{1}{|G|}\right]\cong\widehat{\mathcal{E}ll^*}_G,$$

where completion is with respect to $\varepsilon : \mathcal{E}ll^*_G \to \mathcal{E}ll^*_1 \otimes \mathbb{Z}[1/|G|]$.

The following results complete Devoto's discussion of $\mathcal{E}ll^*_G$.

PROPOSITION 7.4. *The ring $\mathcal{E}ll^*_G$ is a flat $\mathcal{E}ll^*_1$-module.*

Proof: This is a deep result about modular forms of higher level (see [29] relying on [37]). ■

Let $\mathfrak{p}$ be a homogeneous prime ideal of $\mathcal{E}ll^*_G$. We say that $H \subseteq G$ is in the *support* $\mathrm{Supp}(\mathfrak{p})$ of $\mathfrak{p}$ if:

SUPP 1: There exists a homogeneous prime ideal $\mathfrak{p}'$ of $\mathcal{E}ll^*_H$ such that:

$$\mathfrak{p} = i^{*-1}(\mathfrak{p}'),$$

SUPP 2: If $H' \subsetneq H$, then $\mathfrak{p} \neq i^{*-1}(\mathfrak{p}'')$ for any ideal $\mathfrak{p}'' \lhd \mathcal{E}ll^*_{H'}$.

The support is well-defined up to conjugation. Detecting $\mathcal{E}ll^*_G$ by its restrictions to abelian subgroups Corollary 7.1 implies that the support of any $\mathfrak{p} \lhd \mathcal{E}ll^*_G$ is the conjugacy class (H) of some subgroup H equal to the image of some homomorphism of $\mathbb{Z}\times\mathbb{Z}$ into G.

Using Theorem 7.1 we can prove Proposition 7.5 for the prime ideal $\mathfrak{p}$ and subgroup H.

PROPOSITION 7.5. *The following statements are equivalent:*

1. $\mathfrak{p} = i^{*-1}(\mathfrak{p}')$.
2. $\mathrm{Ker}(i^*) \subseteq \mathfrak{p}$.
3. *The localization of $\mathcal{E}ll^*_H$ at $\mathfrak{p}$ is nonzero.*

COROLLARY 7.2. *Let $H \subset G$. If $\mathfrak{p} \lhd \mathcal{E}ll^*_H$ and $K \in \mathrm{Supp}(\mathfrak{p})$, then $K \in \mathrm{Supp}(i^{*-1}\mathfrak{p})$ in G.*

7.2. Example: Metacyclic p-Groups

Let G be presented by:

$$\{A,B : A^{p^{t-1}} = B^p = 1, \qquad B^{-1}AB = A^{1+p^{t-2}}\},$$

so that the E_2-page of the spectral sequence of the extension has

$$E_2^{r,s} = H^r(BC_p, Ell^s(BC_{p^{t-1}})).$$

Previous calculations imply that $Ell^{\text{odd}}(BC_{p^{t-1}}) = 0$ and $Ell^{\text{even}}(BC_{p^{t-1}})$ is p-torsion free. Hence the spectral sequence collapses, $Ell^*(BG)$ is detected by classes lifted from the quotient group $C_p \cong \langle B\rangle$ and invariant classes in $Ell(BC_{p^{t-1}})$. It remains to look at $\mathcal{E}ll^*_G$, the analog of the representation ring. Because the extension defining G is split, the restriction maps $Ell^*(BG)$ onto $Ell^*(BC_p)$; as for ordinary representations, induction from $\langle A\rangle$ to G followed by restriction to $\langle A\rangle$ shows that:

$$\mathcal{E}ll^*_G \to (\mathcal{E}ll^*_{\langle A\rangle})^{\text{inv}}$$

is surjective also. As in the discussion of solvable groups in [13], we must check that various completions used give the same topology. Firstly I-adic completion is an exact functor [13, Sec. 3], provided we fix the group G concerned. Secondly I_H and I_G-topologies coincide on $\mathcal{E}ll^*_H$, and hence on its invariant subset with respect to the action of B. This holds as for K-theory [13, Sec. 6], and it explains our preceding brief discussion of the ideal structure of $\mathcal{E}ll^*_G$.

7.3. Equivariant Elliptic Cohomology of G-Complexes*

The first step in defining an equivariant elliptic genus is to do this for a single G-manifold X. To this end we use a K-theory-valued characteristic class, described in [120] (see also [101]) defined for any real bundle E over X:

$$\theta(E) = \bigotimes_{n=1}^{\infty}[\Lambda_{-q^{2n-1}}(E\otimes\mathbb{C})]\otimes[S_{q^{2n}}(E\otimes\mathbb{C})],$$

where Λ_t (S_t) is defined by the power series $1+\Lambda^1 t+\Lambda^2 t^2+\cdots$ $(1+S^1 t+S^2 t^2+\cdots)$ with coefficients given by exterior and symmetric powers. To allow for G-action, we take values in

$$\mathcal{K}^*_G(X) = \bigotimes_{(g_1,g_2)\in G^{(2)}} \{K^*(X^{g_1g_2})\otimes_{\mathbb{Z}} R\langle g_2\rangle\}[[q^{1/|g_1|}]],$$

*Subsection 7.3 is very condensed, and intended only as an introduction to Devoto's version of equivariant elliptic cohomology. This subject matter is not used in the remaining chapters.

with notation as before; thus $X^{g_1 g_2}$ refers to the fixed point set. Restriction to $X^{g_1 g_2}$ induces a bundle decomposition of $E \otimes \mathbb{C}$ as $\oplus_{j,k} F_{jk}$, where g_1 acts fiberwise on F_{jk} by $e^{2\pi i j/|g_1|}$ and g_2 by $e^{2\pi i k/|g_2|}$. Now write

$$\theta_G(E|_{X^{g_1 g_2}}) = \bigotimes_{j,k} \left[\bigotimes_{s \geqslant 1} (\Lambda_{[\omega^{2k/|g_2|}][-q^{2s-1}]}[F_{jk}]) \otimes \right.$$
$$\left. \bigotimes_{s \geqslant 0} (S_{[\omega^{2k/|g_2|}][q^{2s}]}[F_{jk}]) \right],$$

with $s = (n\,|g_1| + j)/\,|g_1|$ and $R\langle g_2 \rangle = \mathbb{Z}[\omega]$.

Because of our accumulated assumptions, there are Gysin maps:

$$\pi_!^{g_1 g_2} : K^*(X^{g_1 g_2}) \otimes \mathbb{Z}\left[\frac{1}{2}\right] \to K^*(pt) \otimes \mathbb{Z}\left[\frac{1}{2}\right]$$

induced by the map $\pi : X^{g_1 g_2} \to pt$. We can assemble these to give a map $\pi_! : \mathcal{K}_G^*(X) \to \mathcal{K}_G^*$ and define the geometric (twisted) elliptic genus Φ_G by:

$$\Phi_G(X) = \pi_! \left(\frac{\theta_G(T_X)}{\theta_G((\dim TX))} \right) \in \mathcal{K}_G^*.$$

The bundle $\dim TX$ is topologically trivial, but with G-action on a restricted generic fiber corresponding to the actions of the preceding F_{jk}.

This very cumbersome definition arises from combining the nonequivariant definition with the decomposition of a G-manifold X into a union of not necessarily connected strata, fixed under the action of subgroups $H \leqslant G$. (See [40] for the general theory and [33] for the special case of oriented, equivariant cobordism.) Note: The family of subgroups chosen corresponds to the family of rank 2 abelian subgroups used to detect $\mathcal{E}ll_G^*$. Given the Conner–Floyd definition of $\Omega_G^* = \Omega_{-*}^G$ in terms of bordism of oriented manifolds admitting smooth G-actions, we obtain a ring homomorphism from Ω_G^* to $\mathcal{K}_G(pt)$, which is a sum of power series rings of the form $R\langle g_2 \rangle[[q^{1/|g_1|}]]$.

The genus $\Phi_G(X)((g_1,g_2),\tau)$ is obtained by evaluating at the pair (g_2,τ), using the correspondence between characters and representation spaces for the first component. Modularity is built into Witten's definition, so that Φ_G can be thought of as taking values in $\mathcal{E}ll_G^*$.

The definition of Φ_G on Ω_G^* is not quite the same as its definition on coefficients of the equivariant cohomology theory MSO_G^*. For this we need the suspension homomorphism:

$$\Sigma(V) : \Omega_n^G(X,A) \to \Omega_{n+|V|}^G(D(V) \times X, (D(V) \times A) \cup (S(V) \times X))$$

where $D(V)$ $[S(V)]$ is the unit disc (unit sphere) is some representation space V. We fix an infinite dimensional representation space $\mathcal{U}$ containing infinitely many copies of each irreducible representation of G, then partially order the set of finite-dimensional G-subspaces of $\mathcal{U}$ by inclusion.

DEFINITION. The homotopy theoretic equivariant-oriented bordism group of the pair (X,A) equals

$$MSO_n^G(X,A) = \varinjlim_{V \subseteq \mathcal{U}} \Omega^G_{n+|V|}(D(V)\times X, (D(V)\times A)\cup(S(V)\times X)).$$

Restricting attention to coefficients, we must extend Φ_G to a map, or rather family of maps

$$\Phi_G^V : \Omega^G_{n+|V|}(D(V),S(V)) = \widetilde{\Omega}^G_{n+|V|}(\Sigma V) \to \mathcal{E}ll_n^G,$$

where for the usual reasons we confine attention to subgroups $\langle g_1, g_2\rangle$ of G.

Split the representation space V as:

$$V = V_0 \oplus (\oplus_{jk} n_{jk} V_{jk}) = V_0 \oplus V_1,$$

where V_0 is the subspace corresponding to the trivial representation and the V_{jk} are counted out as before. Suspension along the trivial representation is an isomorphism, so that a pair $(M,\partial M)$ mapping into $(D(V),S(V))$ for $\Omega_*^G(D(V),S(V))$ desuspends to a pair $(N,\partial N)$ mapping into $(D(V_1),S(V_1))$. Denote this map by p and note that $N^{g_1,g_2} \in p^{-1}(0)$ because of the absence of fixed points on $D(V_1)$. The tangent bundle splits as:

$$TN|N^{g_1 g_2} = TF \oplus NF,$$

where TF is tangent to the fiber of p, so that (up to modification by a scalar factor) we can use the quotient $\theta_G(TF)/\theta_G(\dim TF)$ to define $\phi_G^V(M,\partial M)$. To bring our definition of cobordism into line with elliptic cohomology, we invert

$$\Delta_{MSO} = [\mathbb{H}P^2]([\mathbb{C}P^2]^2 - [\mathbb{H}P^2])^2 \in MSO_* \subset MSO_*^G.$$

The accompanying factorization of Φ_G now allows us to define our candidate for a G-equivariant elliptic cohomology theory by tensoring:

$$mso_G^*(X,A) = MSO_G^*\left[\frac{1}{2}, \frac{1}{|G|}, \frac{1}{\Delta_{MSO}}\right](X,A)$$

with $\mathcal{E}ll_G^*$ using the genus Φ_G.

REMARK. A simple change of ring argument implies that it is actually enough to take tensor products over MSO_G^*.

THEOREM 7.2. *$\mathcal{E}ll_G^*(X,A)$ previously defined is a stable G-equivariant multiplicative cohomology theory.*

Sketch of proof: Since mso_G^* is constructed to satisfy axioms for a cohomology theory, methods in Chap. 6 imply that for a fixed finite G-CW complex:

$$H \mapsto mso_H^*(X,A)$$

is a Green functor. We already showed that $H \mapsto \mathcal{E}ll_H^*$ is a Green functor. Having checked that the kernel of the equivariant genus behaves well under restriction from G to H, we can combine these two results to show that:

$$H \mapsto \mathcal{E}ll_H^*(X)$$

is again a Green functor. Since coefficients were chosen to contain $1/|G|$, it is enough to check the cohomological properties on rank 2 abelian subgroups. Furthermore by the standard technique of localization at a fixed point set — denoted by the suffix S (see [40]) — we have an isomorphism:

$$\mathcal{E}ll_G^*(X) \to \bigoplus_{\langle g_1,g_2\rangle} [\mathcal{E}ll^*(X^{g_1g_2}) \underset{\mathcal{E}ll^*}{\otimes} \mathcal{E}ll^*_{\langle g_1,g_2\rangle}]^{W\langle g_1,g_2\rangle}_{S\langle g_1,g_2\rangle}.$$

We already listed the result that rings of higher level modular forms $\mathcal{E}ll^*_{\langle g_1,g_2\rangle}$ are flat over $\mathcal{E}ll_1^*$ [Proposition 7.4], from which it follows that the right-hand side satisfies axioms for a cohomology theory. Hence the left-hand side does also. ■

The black box of techniques developed in [5, 6] now allows us to state a general completion result, Theorem 7.3.

THEOREM 7.3. † *If X is a finite $G-CW$-complex and $I = \ker(i^* : \mathcal{E}ll_G^* \to \mathcal{E}ll_1^*)$, there is an isomorphism:*

$$\mathcal{E}ll_G^*(X)_I^\wedge \to \mathcal{E}ll_G^*(EG \times X).$$

Given that we inverted the order of G in the coefficients, as in the case of the main result in Chap. 3, the proof consists of checking that it holds for abelian subgroups $\langle g_1, g_2\rangle$, and then applying Artinian induction. Note: If X = point, the isomorphism is induced by the map $EG \to$ point, so we recover the special case considered in Sec. 7.2.

7.4. Notes

All completion theorems of the type considered in Chap. 7 are motivated by the original work of Atiyah [13], who used the paradigm:

$$\text{cyclic} \xrightarrow[\text{spectral sequence}]{} \underset{\substack{\text{(including}\\ \text{elementary)}}}{\text{solvable}} \xrightarrow[\text{Brauer}]{} \text{arbitrary group } G$$

†Rather than summarize a summary, we refer the motivated reader to the two original papers by Adams et al. [5, 6], rather than to [38].

to establish the relation between $K(BG)$ and the representation ring. Equivariant K-theory allowed a proof of the K_G-analog of Theorem 7.3, using:

$$S^1 \longrightarrow (S^1 \times \cdots \times S^1) \xrightarrow[\text{restriction}]{} U_n \xrightarrow[\substack{\text{holomorphic}\\ \text{induction}}]{} \text{Lie Group } G,$$

which brings compact, but not necessarily connected, groups into the picture. A version of equivariant elliptic cohomology for such compact groups as U_n is introduced in [48] in a flavor very similar to Devoto's; it seems likely that a proof of Theorem 7.3 can be given in this framework along the lines of [16].

We are naturally led to ask about the validity of analogs of theorems in Chap. 7 for other equivariant cohomology theories. The picture is good for many variants of K-theory, for example those defined (at least heuristically) by bundles fibered by vector spaces over $\mathbb{F}_q$ or by finitely-generated projective modules over rings of arithmetic integers (see [99, 88]). Perhaps most interesting is the extreme case of stable cohomotopy. Graeme Segal asked if the natural map:

$$A(G) \to \omega^0(BG)$$

from the Burnside ring can be completed with respect to powers of $I = \ker(A(G) \to A(1))$ to give an isomorphism. After a period of great activity by several authors this was eventually proved by [31], and given the universal status of the Burnside ring, it is a very satisfying result indeed. In the main body of the text we refer to [5, 6], which provide the machinery behind Devoto's results.

Another way to generalize Atiyah's original theorem is to replace the finite group G by a virtual duality, or more particularly, by an arithmetic group. Such a group, $SL_n(\mathbb{Z})$ for example, has a torsion-free subgroup G_1 of finite index such that BG_1 is a finite CW-complex. Alejandro Adem, among others, studied the flat bundle homomorphism:

$$R(G)_{(f)} \to K(BG_1 \underset{Q}{\times} EQ)$$

from finite-dimensional representations of G into the K-group of the classifying space, and his results almost certainly admit analogs for theories where Theorems 7.1 and 7.3 hold (see [7]).

Chapters 3 and 7 are closely related. In the former we reformulated Atiyah's theorem so as to force the definition of the class function ring:

$$Cl(\mathrm{Hom}(\mathbb{Z}_p^n, G), \overline{\mathbb{Q}}_p).$$

Motivated by the isomorphism:

$$R(G) \otimes \mathbb{Q} \to Cl(G, Q(\zeta))^{\mathrm{Gal}(\mathbb{Q}(\zeta), Q)}$$

and making the vital application of Tate–Lubin Theory, [56] proved their version of a completion theorem without benefit of equivariant cohomology. But their result in turn suggested the definition of $\mathcal{E}ll_G^*$ (point), on which Chap. 7 results depend. We return to the starting point if we define $\mathbb{K}^*(x)$ to be a sufficiently large extension of the graded field of fractions of the ring $\mathcal{E}ll_1^*$, where sufficiently large means that we adjoin enough elements to give sense to the following evaluation map. The family of elements x corresponds to the appropriate root of unity ζ in classical representation theory, and it already appeared in our description of $\mathcal{E}ll^*_{\langle g_1, g_2 \rangle}$ in terms of higher level modular forms (Proposition 7.2). To allow for all rank 2 abelian subgroups of G, we note that $\mathbb{K}^*(x)$ is a Galois extension of $\mathbb{K}^*$ with group $GL_2(\mathbb{Z}/|G|)$, and we recall that this group also acts on $G^{(2)}$, as described in Sec. 7.1.

DEFINITION. The *elliptic character ring*:

$$Cl(G^{(2)}, \mathbb{K}^*(x))^{GL_2(\mathbb{Z}/|G|)}$$

is the ring of functions $f : G^{(2)} \to \mathbb{K}^*(x)$ that are invariant with respect to both conjugation on $G^{(2)}$ and the Galois action on $\mathbb{K}^*(x)$.

Evaluating $\vartheta \in \mathcal{E}ll_G^*$ on the pairs $(g_1, g_2) \in G^{(2)}$ gives a holomorphic function that is both Galois and conjugation-invariant, i.e., belongs to $\mathbb{K}^*(x)$ for x chosen suitably. Counting $\mathbb{K}^*$-ranks in both domain and image then shows that evaluation induces an isomorphism:

$$\mathcal{E}ll_G^* \underset{\mathcal{E}ll_1^*}{\otimes} \mathbb{K}^* \cong Cl(G^{(2)}, \mathbb{K}^*(x))^{GL_2(\mathbb{Z}/|G|)}.$$

This result provides the link between [56, 38]

8

Elliptic Objects

The definition of equivariant elliptic cohomology suggests a geometric description of $Ell^*(BG)$ by means of infinite-dimensional bundles over the free loop space LBG. A more general framework is provided by axioms for a topological quantum field theory (TQFT) as presented in [14, 94], with the domain category having additional conformal structure. The natural elegance of this construction is enhanced by the language it provides to describe *moonshines* associated to the groups $\mathbb{M}$, Co_0, and M_{24}. We introduced this in Chap. 4, where we saw that the action of the Hecke algebra on modular forms leads to the construction of a natural Thompson series for the smallest of these groups. Hecke operators admit an interpretation as cohomology operations inside elliptic cohomology, showing that the existence of at least one moonshine can be explained cohomologically. The same may well be true for Co_0 and $\mathbb{M}$, providing one good reason for advancing calculations in Chap. 5. We cover this material in Secs. 8.1–8.3. The remaining sections are more tentative — using [29], the definition of an elliptic object over BG can be generalized to more general spaces X. Brylinski [29] assumes that the base space X is a 1-connected manifold; our putative generalization shows that the moonshine-like construction for BG extends to $\pi_1 X$. This more general definition of an elliptic object — formulated jointly as part of a continuing program in [18] — covers many of the existing examples. At best as results in Chap. 6 show, our elliptic objects detect a part of $Ell^*(X)$, in particular we have nothing to say about connections, which [101] suggests as an integral part of the structure. One reason for downplaying these in the case of BG is that our definition has a built-in assumption of flatness.

8.1. Quantum Field Theories

A d-dimensional topological quantum field theory over $\mathbb{C}$ involves the following data:

1. A possibly infinite dimensional state space H_Σ defined over $\mathbb{C}$ and associated with each smooth, closed, oriented d-dimensional manifold Σ.

2. An operator $U_M : H_{\Sigma_0} \to H_{\Sigma_1}$ associated with each $(d+1)$-dimensional manifold M cobounding $\overline{\Sigma}_0 \sqcup \Sigma_1$. Here $\overline{\Sigma}_0$ denotes Σ_0 with the opposite orientation; when H is infinite-dimensional, the pair (H, U) satisfies some finiteness assumption, e.g., U is of trace class. The map $\mathcal{Z}$ just described satisfies the following rules:

   a. $\mathcal{Z}$ is *functorial* with respect to orientation preserving diffeomorphisms of Σ and M.

   b. $\mathcal{Z}$ is *involuntary*, i.e., $H_{\overline{\Sigma}} = (H_\Sigma)^*$, the dual space.

   c. $\mathcal{Z}$ is *multiplicative*, i.e., (on spaces) $H_{\Sigma_0 \cup \Sigma_1} = H_{\Sigma_0} \otimes H_{\Sigma_1}$ and (on maps) $U_{M'} \circ U_M = U_{M' \cup M}$.

The composition rule for operators is referred to as transitivity with respect to gluing cobordisms.

The description just given is supposed to be indicative, and in presenting it we follow [14]. Greater emphasis on functoriality is given by some authors, for example [94], but too tight a definition imposes restrictions we wish to avoid, including the finite dimensionality of H_Σ. The need to allow infinite-dimensional spaces and representations is one reason for allowing scalar multiples of maps and projective as opposed to linear representations. In passing we note that Quinn's definition of a domain category, and its associated cobordism category, and the multiplicative functoriality of $\mathcal{Z}$ makes rule (b) redundant; i.e., involution is determined by the remaining structure.

In many theories of physical interest, the manifold pair (M, Σ) carries additional geometric structure, and (c) (for maps) only holds up to a scalar factor. For example [101] introduces a *conformal* as opposed to a *topological* field theory by projectivizing U_M as before, and also taking M to be a bounded Riemann surface. A cobordism from Σ_0 to Σ_1 then consists of a pair (M, α), where α is an isomorphism between ∂M and $\Sigma_1 - \Sigma_0$. Two such pairs (M, α) and (M', α') are identified if they are isomorphic.

It is convenient to refer to M^{d+1} as a space–time, and Σ_0 (Σ_1) as an incoming (outgoing) boundary.

Note: Multiplicativity is a very strong condition, it asserts for example that $\mathcal{Z}(M)$ is independent of how we cut M in half along some codimension 1 submanifold Σ.

Allowing if necessary for conformality, we lose nothing essential if we impose further conditions:

d.

$$
\begin{aligned}
H_\varnothing &= \mathbb{C}(\varnothing = \text{empty } \Sigma^d),\\
U_\varnothing &= 1(\varnothing = \text{empty } M^{d+1}),\\
U_{\Sigma\times[0,1]} &= Id.
\end{aligned}
$$

As elementary consequences of the axioms, we have:

e. If $\partial M^{d+1} = \varnothing, U_M \in \mathbb{C}$. This implies that our theory provides a numerical invariant for $(d+1)$-manifolds.

f. $U_{\Sigma\times S^1} =$ "trace" $(Id|H_\Sigma) =$ "Dim" H_Σ (suitably defined).

More generally if we form the mapping torus Σ_f for some diffeomorphism $f:\Sigma\to\Sigma$, then:

$$U_{\Sigma_f} = \text{"trace"}\ (f_*|H_\Sigma).$$

One further general item needs to be mentioned — the effect of orientation reversal on $\mathcal{Z}(M)$ when M is a closed $(d+1)$-manifold. To explain this, we assume that the $\mathbb{C}$-vector spaces $\mathcal{Z}(\Sigma)$ admit nondegenerate Hermitian forms with respect to conjugation in $\mathbb{C}$ and impose the additional condition:

g. $\mathcal{Z}(\overline{M}) = \overline{\mathcal{Z}(M)}$.

This is compatible with Rule b on $\mathcal{Z}(\overline{M})$, since in the presence of a Hermitian form we can identify $\mathcal{Z}(\overline{\Sigma})$ and $\overline{\mathcal{Z}(\Sigma)} \cong \mathcal{Z}(\Sigma)^*$. If M cobounds $\overline{\Sigma}_0 \cup \Sigma_1, \mathcal{Z}(M)$ can be interpreted as a linear transformation of Hermitian spaces, and in light of Rule g, reversing of the orientation of M replaces $\mathcal{Z}(M)$ by its adjoint mapping. Thus over $\mathbb{C}$ we expect $\mathcal{Z}(M)$ to be sensitive to orientation, and if DM denotes the double of M along the boundary $\partial M = \Sigma$, then $\mathcal{Z}(DM) = \|\mathcal{Z}(M)\|^2$.

Rule e is important in considering moonshine, since it implies that the conformal invariance properties of representation modules H_Σ are reflected in the modularity properties of the graded trace. We expand on this point, taking Segal's conformal theory with $d = 1$ as our example. Invariance under diffeomorphisms implies that the Hilbert space $V = \mathcal{Z}(S^1)$ has the structure of a $\mathrm{Diff}^+(S^1)$-representation space, on which we impose the following restrictions:

1. Under the action of the infinitesimal generator L_0 of the rotation subgroup $SO_2 \hookrightarrow \mathrm{Diff}^+ S^1$, the representation space V decomposes as a direct sum:

$$V = \bigoplus_{n\in\frac{1}{m}\mathbb{Z}} V_n$$

with $\dim V_n < \infty$ for each n.

2. $V_n = 0$ for $n \leqslant n_0$, $n_0 \in \mathbb{Z}$.

Condition (2) states that V is a representation of almost positive energy. (This follows a suggestion in [92, p. 172] to replace $V_n = 0$ for $n < 0$ as the positive energy condition with $n \leqslant n_0$. The two are essentially equivalent, since we can always multiply the action of SO_2 on V by some character of SO_2.) If we think of positive energy as a holomorphic condition, then almost-positive energy is meromorphic. Corresponding to a cobordism M between Σ_0 and Σ_1, we have a linear operator $\mathcal{Z}(M) : \overline{V}_0 \to V_1$, which admits a graded trace, that we can write as a power series:

$$\sum_{\substack{n \in \frac{1}{m}\mathbb{Z} \\ n \geqslant n_0}} \dim_{\mathbb{C}}(V_n) q^n, \qquad q = e^{2\pi i \tau}.$$

At least when $m = 1$ and $n_0 = 0$, the modularity of this function is forced on us by Segal's *contraction property*, which states that if M is a cobordism between Σ and itself, then the (graded) trace of the operator $\mathcal{Z}(M) : \mathcal{Z}(\Sigma) \to \mathcal{Z}(\Sigma)$ depends only on the closed manifold obtained by gluing the two copies of Σ to each other. But we already saw that this is part of the structure of a TQFT, due to the multiplicative property. We observe in particular that if M_τ is the annulus:

$$\{z \in \mathbb{C} \,|\, |e^{i\tau}| \leqslant |z| \leqslant 1\},$$

then identifying boundary components gives the same closed manifold for both τ and $\tau' = -1/\tau$. It follows that the preceding trace function is invariant under the map $\tau \mapsto -1/\tau$. This is a major step in proving modularity for level 1 and weight zero, since a special linear map:

$$\tau \mapsto \frac{a\tau + b}{c\tau + d} \qquad (c \neq 0)$$

of H_1 to itself can be decomposed as:

$$\tau \mapsto \frac{-1}{c^2\left(\tau + \frac{d}{c}\right)} + \frac{a}{c},,$$

i.e., as the product of a translation, inversion, dilation, and another translation. The decomposition when $c = 0$ is even easier.

Remark. We can complicate the situation slightly by requiring M to have a fixed spin structure. This amounts to requiring our linear transformations to fix a preferred point in H_1 (compare the discussion in Sec. 2.1) and it reduces the invariance subgroup to $\Gamma_0(2)$.

This can be related to K-theory if we think of a vector bundle (with connection) as a functor from the category SDiff_X^{0+1} of points and smooth paths in X into finite-dimensional vector spaces. Introducing a parameter space X into Segal's cobordism category, we similarly obtain Con_X^{1+2}, where the objects are points $s : \Sigma \to X$ in the free loop space LX of X; and morphisms are triples (M, α, m), where $m : M \to X$ makes the following diagram commute:

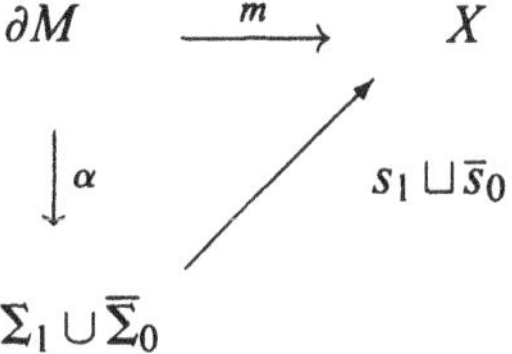

This means that the restriction of M to the boundary of the Riemann surface M is compatible with identifying α on this boundary with loops Σ_0 and Σ_1.

DEFINITION. An *elliptic object* of weight k on X is a functor:

$$E : \mathrm{Con}_X^{1+2} \longrightarrow \{\text{weight } k \text{ representations of } \mathrm{Diff}^+ S^1 \text{ of finite type}\}.$$

Our discussion shows that meromorphicity and the contraction condition are built into this definition. An elliptic object thus consists of an infinite-dimensional vector bundle over the loop space LX, equivariant under $\mathrm{Diff}^+ S^1$, and some type of connection. We have least to say about this last part of the structure; however in the special case of LBG the connection turns out to be *flat*, i.e., the holonomy associated with translation across the surface M is expressed in terms of a homomorphism:

$$\pi_1(M) \to G.$$

At least when M is a torus, this fits in well with the preferred status of rank 2 abelian subgroups of G.

8.2. Free Loop Space LBG

Let G be a finite group. It is well-known that BG and EG can be expressed as unions of finite-dimensional smooth manifolds. For example if $|G| = m$, by considering the permutation action of G on itself we can embed G in either O_m or U_m then take EG to be a union of Stiefel manifolds, each fibering over the appropriate Grassmann orbit space. If $b_0 \in BG$ is some chosen base point and $f : S^1 \to BG$ is a point in the free loop space LBG, we distinguish loops based at b_0 by $f_0 \in \Omega_{b_0} BG$.

PROPOSITION 8.1. *The connected components of LBG are indexed by the set of conjugacy classes of elements $g \in G$, and the component corresponding to the class $[g]$ is homotopy-equivalent to $BC_G(g)$.*

Proof: For each $g \in G$ define a subspace of the space of all smooth paths in the universal space EG by:

$$L_gEG = \{p : \mathbb{R} \to EG : p(t+1) = gp(t) \text{ for all } t \in \mathbb{R}\},$$

and let

$$L_GEG = \bigsqcup_{g \in G} L_gEG.$$

An element $p \in L_gEG$ is determined by its values on the unit interval $[0,1]$; we can identify L_gEG with the space of suitably smooth maps $\{p : [0,1] \to EG : p(1) = gp(0)\}$. There is a smooth left action of G on L_GEG given by:

$$(gp)(t) = g(p(t)),$$

under which $h \in G$ maps L_gEG to $L_{hgh^{-1}}EG$. Hence L_gEG admits an action by the centralizer $C_G(g)$. Note: The quotient space L_GEG/G can be identified with LBG and each space L_gEG is contractible. To prove the latter, let $H : [0,1] \times EG \to EG$ be a (based) contraction for which $H(0,p) = p$ and $H(1,p) = e_0$, a base point for EG over b_0; define $H_g : [0,1] \times L_gEG \to L_gEG$ by:

$$H_g(t,p)(s) = \begin{cases} H(t,p(s)) & \text{if } 0 \leqslant s+t \leqslant 1 \\ H\left(1-s, gH\left(\frac{s+t-1}{s}, g^{-1}p(s)\right)\right) & \text{if } 1 \leqslant s+t \leqslant 2. \end{cases}$$

The G-universality of EG implies that there are G-equivariant maps $L_gEG \to EG \times \{g\}$ and thus a G-equivariant map:

$$L_GEG \to EG \times G$$

whose right G-action on the target space is given by:

$$(e,g)h = (he, hgh^{-1}).$$

Taking quotients we see that:

$$LBG \cong L_GEG/G \simeq EG \underset{G}{\times} G^{(c)},$$

where superscript c specifies that the action on G is given by conjugation. The proposition now follows. Note: The component corresponding to $[g]$ is homotopy-equivalent to $BC_G(g)$ by our earlier remark on the action of G on paths. In more detail we have the pullback diagrams:

$$\begin{array}{ccc} L_gEG \underset{C}{\times} G & \longrightarrow & EG \\ \downarrow & & \downarrow \\ L_gBG & \longrightarrow & BG \end{array} \qquad \begin{array}{ccc} L_gEG \underset{C}{\times} F & \longrightarrow & EG \underset{G}{\times} F \\ \downarrow & & \downarrow \\ L_gBG & \longrightarrow & BG \end{array},$$

where F denotes any G-space. These show that L_gBG has the properties required of the classifying space for $C_G(g)$. ■

The preceding discussion extends to the double-loop space:

$$L^2BG = \mathrm{Map}(S^1 \times S^1, BG) = L(LBG),$$

with elements F restricting to two loops, $z \mapsto F(z,-)$ and $z \mapsto F(-,z)$. Analogs of spaces L_gEG are the spaces:

$$\begin{aligned} L^2_{g,h}EG = \{F : \mathbb{R}^2 \to EG : F(x+1,y) &= gF(x,y) \\ F(x,y+1) &= hF(x,y) \qquad \text{for all } x,y \in \mathbb{R}\}, \end{aligned}$$

defined for all pairs $(g,h) \in G^{(2)}$ with $h \in C_G(g)$.

We can combine these spaces into the disjoint union over all such commuting pairs:

$$L^2_GEG = \sqcup_{g,h} L^2_{g,h}EG.$$

The group G acts as before, and k maps $L^2_{g,h}EG$ onto $L^2_{kgk^{-1},khk^{-1}}EG$; thus $L^2_{g,h}EG$ itself admits a free action by the subgroup $C_G(g,h) = C_G(h) \cap C_G(g)$. A similar argument shows that:

$$L^2_{g,h}EG/C_G(g,h) \cong BC_G(g,h).$$

Combining Proposition 8.1 and the definition of an elliptic object, we see that an elliptic object over LBG consists of a family of graded infinite-dimensional bundles over the classifying spaces $BC_G(g)$ as g runs through a family of representatives for the conjugacy classes in G. Each summand of the graded bundle is finite-dimensional, and it satisfies a modularity condition because of the contraction property. By Atiyah's completion theorem in K-theory, flat bundles are dense in $K(BC_G(g))$; therefore elliptic objects associated to representations of centralizers of elements $g \in G$ play a privileged role. Such families of representations arise naturally in various mathematical contexts, for example see [46, 23], and provide a motivation for choosing coefficients in equivariant elliptic cohomology.

DEFINITION. Let G be a finite group. A *G-elliptic system* is a family of graded $C_G(g)$-representation spaces $\{H_g : g \in G\}$, such that the function $\vartheta(g,h,\tau)$ defined on $G^{(2)} \times H_1$ by:

$$\vartheta(g,h,\tau) = \text{graded trace of } h \text{ on } H_g,$$

with $H_g = \bigoplus_{n \geqslant n_0} H_{g,n} q^{n/D_g} \quad (q = e^{2\pi i \tau}, D_g$ dependent only on $g)$, satisfies

1. $\vartheta(g,h,\tau)$ is meromorphic on $H_1 \cup \{\infty\}$.
2. ϑ is invariant under conjugation of the commuting pair (g,h).
3. for some weight k and some twisting Dirichlet character $\sigma = \sigma(A;g,h)$:

$$\vartheta(g,h,\tau) = \sigma(c\tau+d)^{-k} \quad \vartheta(g^d h^{-c}, g^{-b} h^a, A\tau).$$

Here as always $A = \begin{pmatrix} a & b \\ c & d \end{pmatrix} \in SL_2(\mathbb{Z})$.

In Chap. 4 we defined such a family of functions for $G = M_{24}$ and $g = 1$ [so that $C_G(g) = G$]. In Sec. 8.3 we complete this definition for all rank 2 abelian subgroups, obtaining a two-variable Thompson series with modular characters. For the moment we note that G-elliptic systems are closely related to elements in the ring $\mathcal{E}ll_G^*$, since the character of a representation satisfies the Devoto condition Dev-3. As we noted when defining $\mathcal{E}ll_G^*$ this step is not quite reversible, since we may have to invert the order of G in the coefficients of $R(C_G(g))$. Furthermore the matrix A now belongs to the full modular group rather than the subgroup $\Gamma_0(2)$. This is just what we expect, since in the interesting cases, $|G|$ is even and we work at Level 1 rather than Level 2. In the language of elliptic cohomology, we are looking at a theory with coefficients in:

$$\mathbb{Z}\left[\frac{1}{6}\right][g_1, g_2, \Delta^{-1}]$$

rather than:

$$\mathbb{Z}\left[\frac{1}{2}\right][\delta, \varepsilon, \Delta^{-1}],$$

but with this change, a metatheorem states that the I-adic completion $(\mathcal{E}ll_G^*)^\wedge$ is still isomorphic to $Ell^*(BG)$, subject to restrictions on the order and subgroup structure of G.

8.3. Elliptic System for M_{24}

Chapter 4 posed the problem of finding one-dimensional spaces of eigenforms for the Hecke algebra by exploiting restrictions on weight and level. The solution was closely related to the cycle decomposition of conjugacy classes of elements in the Mathieu group M_{24}, and using bare hands, generating forms can be written in terms of the Dedekind η-function. More precisely we consider the forms:

$$\vartheta(1,h,\tau) = \eta_h(\tau) = \eta(\tau)^{j_1}\eta(2t)^{j_2}\cdots\eta(s\tau)^{j_s} \qquad j_1+2j_2+\cdots+sj_s = 24.$$

Taken together as h runs through the conjugacy classes in M_{24}, these forms coincide with the graded character of the infinite-dimensional representation:

$$\Theta = q\prod_{k=1}^{\infty}\sum_{r=0}^{\infty}(-1)^r \wedge^r(V)q^{kr},$$

where $\wedge^r(V)$ is the r exterior power of the permutation representation of M_{24} in S_{24}. Identification of form and character follows by inspecting the characteristic polynomial of the permutation matrix representing h. We refer to Θ as a Thompson series for M_{24} or more suggestively for the centralizer of identity 1 in M_{24}. We obtain an elliptic system by constructing similar series for other centralizers $C_G(g)$. First we need some more notation.

The pair (g,h) of commuting elements is said to be *rational* if h acts rationally on each of the g-eigenspaces of $V\underset{Q}{\otimes}\mathbb{C}$. If g has order $|g|$, then on the:

$$\exp(2\pi\ell i/|g|)\text{-eigenspace}$$

we can again look at the (restricted) characteristic polynomial of h and write the modular form as:

$$\vartheta(g,h;\tau) = \prod_{\ell||g|}\prod_{d|\ell}\prod_{r_\ell||h|}\eta(r_\ell\tau/d)^{j(r_\ell)\mu(|g|/d)},$$

where μ denotes the Möbius function. When $|g|=1$, we recover the earlier expression for $\vartheta(1,h;\tau)$. Not all pairs (g,h) are rational, and for these a more complicated formula for $\vartheta(g,h;\tau)$ must be used. For the sake of clarity, we omit this, referring the reader to [80, Formulae 3.7 and 4.20]. We also refer to the cusp form:

$$p(\tau) = \prod_r \eta(r\tau)^{j_r}$$

with r increasing and j_r positive, as *primitive* if the corresponding partition of 24 is *balanced*. This means

1. $\sum_r rj_r = 24$.

2. The first cycle length r_1 divides all those r_i that follow, $1 \leqslant i \leqslant s$.

3. If $N = r_1 r_s$, then for $i \geqslant 1$, $N = r_i r_{s+1-i}$.

4. $j_{r(i)} = j_{r(s+1-i)}$.

The integer N in Condition 3 is the *balancing number* of the partition. Inspecting conjugacy classes, we see that the 21 forms $\vartheta(1,h;\tau)$ are all primitive; a side product of the construction we are outlining is that most of the remaining primitive forms correspond to subgroups of rank 2. In particular the two anomalous partitions $(4^2 8^2)$ and $(3^2 9^2)$ from Chap. 4 fit into the elliptic system. We also write

- N_g = balancing number of g in M_{24}.
- $N_{(g,h)} = N_g N_h$.
- $N_A = \min\{N_{(g,h)} : \langle g,h\rangle = A\}$.
- $\mu(g,h;\tau) = \vartheta(g,h;N_g\tau)$.

Replacing ϑ by μ can be thought of as a normalization to obtain the correct level for an element in $\mathcal{E}ll^*_{M_{24}}$.

By listing all abelian subgroups in M_{24} of rank $\leqslant 2$ and by using the preceding formula for $\vartheta(h,g;\tau)$ and its nonrational generalization, G. Mason [81, 80] proved a sequence of results, which include the following:

1. To each abelian subgroup A with at most two generators, there is associated a primitive cusp form $p_A(\tau) = p(\tau)$. If $A = \langle g,h\rangle$, then $\mu(g,h;\tau) = p_A(\tau)$ if and only if $N_{(g,h)} = N_A$ = level of $p(t)$ as a primitive cusp form. In general if:

$$p(\tau) = \sum_{n=1}^{\infty} a_n q^n,$$

there is a root of unity λ such that:

$$\mu(g,h;\tau) = \sum_{n=1}^{\infty} b_n q^n$$

with either $b_n = 0$ or $b_n = \lambda^{n-1} a_n$.

2. For a fixed conjugacy class $[g]$, the forms $\mu(g,h;\tau)$ provide the character of a Thompson series for the centralizer of g. This character property shows that the elliptic system can be thought of as living in the coefficients of equivariant elliptic cohomology. [The definition of the forms shows that the integrality condition (Dev-2) is satisfied, and the Galois condition (Dev-3) holds automatically.] Note: The weight varies with the form; hence as yet, we have only constructed an inhomogeneous element of $\mathcal{E}ll^*_{M_{24}}$. Mason's proof depends on applying his general formula to the pairs (g,h) contained in an abelian subgroup of M_{24} generated by at most two elements. We already described what happens to cyclic groups in Chap. 4; most of the remaining information is found in the *Atlas*. We obtain groups isomorphic to $C_2 \times C_2$, $C_2 \times C_4$, $C_4 \times C_4$, $C_2 \times C_8$, $C_2 \times C_6$, $C_2 \times C_{10}$, and $C_3 \times C_3$. For each of these it is then possible to list the pairs they contain in terms of the pA, qB, rC notation of the *Atlas*. See [84] for a table of the resulting forms; the reader is warned that full information is given only for the forms $\mu(g,h;\tau)$ associated with rational pairs. Given that a rigorous proof of the completion theorem exists at present only for groups of odd order, we note that there are two types of subgroup isomorphic to $C_3 \times C_3$. Restricting attention to a generating pair of elements, we have

$$
\begin{array}{lll}
C_3 \times C_3A & \vartheta(3A,3A;\tau) = \eta(3\tau)^8 & N = 9 \\
C_3 \times C_3B & \vartheta(3B,3A;\tau) = \eta(3\tau)^2\eta(9\tau)^2 & N = 27.
\end{array}
$$

REMARKS. There are simple relations between the forms $\mu(g,h;\tau)$ and $\mu(h^{-1},g;\tau)$ and also between $\mu(g,h;\tau)$ and $\mu(g,g^r h;\tau)$, where $r = N/Q$ for some divisor Q. These are useful in discussing levels and comparing coefficients of $p(\tau)$ and $\mu(g,h;\tau)$. If we list all possible partitions corresponding to primitive forms, we obtain the 21 associated with conjugacy classes in M_{24}, and $(4^2 8^2)(3^2 9^2)$ (see the preceding discussion) $(8\ 16)(4\ 20)(2^3 6^3)$ (associated with commuting pairs) and $(2\ 2\ 2)(6\ 18)$. At least one of the last two matches a conjugacy class in the almost-simple group Co_0.

It is possible to homogenize the forms μ to obtain an elliptic system of constant weight equal to zero by means of θ-functions. As in the case under consideration, we assume that the representation space V is a 24-dimensional lattice admitting an integral quadratic form Q that takes even values. The associated *theta function* is defined by:

$$
\theta_V = \sum_{x \in V} q^{Q(x)/2} = \sum_{n=0}^{\infty} a_n q^n,
$$

with a_n = number of elements in L for which $Q(x) = 2n$. As in the case of the forms η_g, we can associate θ_V with a Thompson series:

$$\Theta'_G = \sum_{n \geqslant 0} A_n q^n,$$

where A_n is the permutation representation of G on the set:

$$V_n = \{x \in V : Q(x) = 2n\}.$$

Contained in V are the fixed sublattices V^g for each conjugacy class of elements g; it is possible to show that both the level and weight of the character θ_{V^g} coincide with those of η_g; i.e., that the quotients:

$$\{\theta_{V^g}/\eta_g : [g] \in G\}$$

are the forms we are seeing. Once more the Galois condition (Dev-3) holds because the forms are genuine characters; the integrality condition (Dev-2) holds because the apparent quotient character corresponds to the representation:

$$\Theta'_G \left(q^{-1} \prod_{k=1}^{\infty} \sum_{r=1}^{\infty} S^r(V) q^{kr} \right).$$

This last graded representation is almost of the kind arising as moonshine, i.e., satisfying a genus zero condition. With further modification of the character functions, this also can be satisfied (see [42]). The genus zero condition states that the modular function (modular form of weight zero) either vanishes identically, or it is a Hauptmodul; i.e., it generates the field of functions on the compactified orbit space $\Gamma \backslash H_1$, where Γ is the invariance subgroup. Table 8.1 lists Mason's results, confining attention to the pairs $(1,h)$. For graded characters satisfying the genus zero condition, this is not too serious a restriction, since for pairs (g,h) with $g \neq 1$, the forms are trivial unless the orders of g and h both equal 3 (see [84, Theorem A]). The notation θ_h/η_h was already explained; an entry such as:

$$(2^3) + 1^6 3^6 / 2^6 6^6$$

should read as:

$$8 + (\eta(\tau)\eta(3\tau)/\eta(2\tau)\eta(6\tau))^6.$$

The character of the quotient of the Thompson series Θ'_G and Θ_G survives under genus zero restriction only for elements of odd order. Does this fact have anything to do with the result, explained in Chap. 9, of incorporating the prime 2 into the definition of elliptic cohomology? The integer N equals the level of the form η_h, obtained by taking the product of the longest and shortest cycles given by

Table 8.1. Elliptic System for M_{24}

Pair$(1,h)$	N	Group	Formula
$(1,1^{24})$	1	$\Gamma_0(1)$	θ_h/η_h
$(1,1^8 2^8)$	2	$\Gamma_0(2)-$	$(2^7)+1^{24}/2^{24}$
$(1,2^{12})$	4	$\Gamma_0(2)-$	$1^{24}/2^{24}$
$(1,1^6 3^6)$	3	$\Gamma_0(3)+$	θ_h/η_h
$(1,3^8)$	9	$\Gamma_0(3\|3)$	θ_h/η_h
$(1,2^4 4^4)$	8	$\Gamma_0(4)-$	$1^8/4^8$
$(1,1^4 2^2 4^4)$	4	$\Gamma_0(4)-$	$(2^5)+1^8/4^8$
$(1,4^6)$	16	$\Gamma_0(4\|2)-$	$2^{12}/4^{12}$
$(1,1^4 5^4)$	5	$\Gamma_0(5)+$	θ_h/η_h
$(1,1^2 2^2 3^2 6^2)$	6	$\Gamma_0(6)+3$	$(2^3)+1^6 3^6/2^6 6^6$
$(1,6^4)$	36	$\Gamma(6\|3)$	$3^8/6^8$
$(1,1^3 7^3)$	7	$\Gamma_0(7)+$	θ_h/η_h
$(1,1^2\cdot 2\cdot 4\cdot 8^2)$	8	$\Gamma_0(8)-$	$(2^3)+1^4 4^2/2^2 8^4$
$(1,2^2 10^2)$	20	$\Gamma_0(10)+5$	$1^4 5^4/2^4 10^4$
$(1,1^2 11^2)$	11	$\Gamma_0(11)+$	θ_h/η_h
$(1,2\cdot 4\cdot 6\cdot 12)$	24	$\Gamma_0(12)+3$	$1^2 3^2/4^2 12^2$
$(1,12^2)$	144	$\Gamma_0(12\|6)$	$6^4/12^4$
$(1,1\cdot 2\cdot 7\cdot 14)$	14	$\Gamma_0(14)+7$	$(2)+1^3 7^3/2^3 14^3$
$(1,1\cdot 3\cdot 5\cdot 15)$	15	$\Gamma_0(15)+$	θ_h/η_h
$(1,3\cdot 21)$	63	$\Gamma_0(21\|3)+$	θ_h/η_h
$(1,1\cdot 23)$	23	$\Gamma_0(23)+$	θ_h/η_h

the embedding in S_{24} (see Chap. 4). Given the identification of a summand of $\mathcal{E}ll^*_{\langle 1,h\rangle}$ with a ring of higher level modular forms in Proposition 7.2, the group Γ_0 is significant. The notation is introduced by J. Conway and S. Norton in their original paper on Moonshine [36]; in particular:

$$\Gamma_0(n|a) = \begin{pmatrix} a^{-1} & 0 \\ 0 & 1 \end{pmatrix} \Gamma_0(n/a) \begin{pmatrix} a & 0 \\ 0 & 1 \end{pmatrix},$$

and + odd refers to the adjunction of certain involutions to the group. Inspection shows that the level of the group is better adapted to the order of h in the pair $(1,h)$ than is the level of the unmodified form η_h.

8.4. Bundles over Arbitrary Loop Spaces LX

In this Sec. 8.4 we formulate a definition of a Virasoro equivariant vector bundle over an arbitrary loop space LX. This simultaneously generalizes both the definition of an elliptic object over BG and the definition of a Virasoro bundle over the loop space of a simply connected manifold used by J. L. Brylinski in [29]. We warn the reader that our definition is provisional in that it is formulated to fit two distinct examples in the literature, and as the results of Chaps. 6 and 7 suggest, it is certainly capable of refinement.

In what follows let the field $\mathbb{K}$ be the real or complex numbers and write $\mathbb{T}$ for the unit circle S^1 (based at 1). Let $\mathfrak{diff}_{\mathbb{K}}$ denote the Lie algebra $\mathbb{K}[z,z^{-1}](d/dz)$ consisting of elements of the form:

$$f(z)\frac{d}{dz}, \qquad f(z) \in \mathbb{K}[z,z^{-1}]$$

with Lie bracket

$$\left[f(z)\frac{d}{dz}, g(z)\frac{d}{dz}\right] = (f(z)g'(z) - g(z)f'(z))\frac{d}{dz}.$$

For us the Virasoro algebra $\mathfrak{vir}_{\mathbb{K}}$ is a universal central extension $\mathfrak{diff}_{\mathbb{K}} \oplus \mathbb{K}$ with basis elements $L_n = (-z^{n+1}d/dz, 0)$ for $n \in \mathbb{Z}$ and $c = (0,1)$ endowed with the Lie bracket satisfying

$$[L_m, L_n] = (m-n)L_{m+n} + \frac{m^3 - m}{12}c$$

and having all other brackets of basis elements equal to zero. Projection onto $\mathfrak{diff}_{\mathbb{K}}$ gives rise to a short exact sequence of Lie algebras:

$$0 \to \mathbb{K} \to \mathfrak{vir}_{\mathbb{K}} \to \mathfrak{diff}_{\mathbb{K}} \to 0.$$

REMARKS.

1. We introduce the central extension to allow for projective as well as ordinary representations. Projectivity was built into the definition of a TQFT in Sec. 8.1, and it is forced on us in physically motivated examples.

2. The factor $m^3 - m$ is again forced on us by the cocycle condition defining the central extension:

   $$\alpha : \mathfrak{diff}_{\mathbb{K}} \times \mathfrak{diff}_{\mathbb{K}} \to \mathbb{K}.$$

   If $\alpha_m = \alpha(L_m, L_{-m})$, then $\alpha_{-m} = -\alpha_m$ and

   $$(m+2n)\alpha_m - (2m+n)\alpha_n = (m-n)\alpha_{m+n}.$$

   The general solution of this recurrence relation is $\alpha_m = \lambda m^3 + \mu m$, and the value of μ is unimportant, given that the particular solution $\alpha_m = m$ is a coboundary. We take $\lambda = -\mu = c/12$. The particular cocycle α is characterized by the fact that it is invariant under rotation and vanishes on $\mathfrak{sl}_2$.

Now suppose that Y is a manifold, which may be infinite-dimensional, and $\mathfrak{diff}_{\mathbb{K}}$ is represented in $\mathrm{Vect}_{\mathbb{K}}(Y)$, the Lie algebra of smooth vector fields on Y taking values in $\mathbb{K}$. If $\xi \to Y$ is a smooth $\mathbb{K}$-vector bundle over Y, then an action of the Virasoro algebra $\mathfrak{vir}_{\mathbb{K}}$ on ξ is an action on sections of ξ compatible with the action of $\mathfrak{diff}_{\mathbb{K}}$. This means that for each open subset U of the base space Y, we have an action of the Lie algebra on the space of smooth sections $\Gamma(U,\xi)$ over U, such that:

- The action is compatible with restriction maps $\Gamma(U_1,\xi) \to \Gamma(U_2,\xi)$ for open sets $U_2 \subseteq U_1 \subseteq Y$.
- For U open in Y, f a smooth function on U, σ a smooth section of ξ over U, and v an element of the Lie algebra:
$$v(f\cdot\sigma) = v(f)\cdot\sigma + f\cdot v(\sigma).$$

Note: These conditions give the infinitesimal analog of a Lie group action on Y. In fact the first example we have in mind is $Y = LM$, where M is a smooth, simply connected, finite-dimensional manifold, and the group acting is $\mathrm{Diff}^+(S^1)$. Brylinski then defines an *admissible* bundle ξ to be one whose restriction to the fixed point subset of constant loops decomposes as a direct sum of finite-dimensional bundles under the action of the infinitesimal generator L_0 of SO_2. Furthermore the indexing set for this decomposition is bounded below. We include this admissibility condition in our more general definition.

As in the special case $X = BG$, the free loop space LX has components indexed by conjugacy classes in the fundamental group $\Pi = \pi_1(X,x_0)$, where x_0 is the chosen base point. Let $\widetilde{X}$ be the universal covering space of X; if $g \in \Pi$, write

$$L_gX = \{p : \mathbb{R} \to \widetilde{X} : p(t+1) = gp(t) \forall t \in \mathbb{R}\},$$
$$L_\Pi X = \bigsqcup_{[g]} L_gX.$$

As for BG we see that $L_gX/C_\Pi(g)$ can be identified with the component $L_{[g]}X$, consisting of loops that lift to paths in $\widetilde{X}$ having holonomy in the conjugacy class $[g]$. A similar construction for the double-loop space L^2X gives spaces $P^2_{g,h}(X)$, defined for all $g,h \in \Pi$ with $h \in C_\Pi(g)$, with union $P^2(X)$. This time the intersection:

$$C_\Pi(g,h) = C_\Pi(g) \cap C_\Pi(h)$$

acts freely on $L^2_{g,h}(X)$.

Now let $\zeta \to LX$ be a vector bundle locally modeled on a Hilbert space $\mathcal{H}$ over the field $\mathbb{K}$. As and when it is necessary, we may need to restrict $\mathcal{H}$, for example to ensure the existence of partitions of unity (see [72]). We also assume that ζ is associated to a principal bundle $\mathcal{Q} \to LX$ with structural group $\mathcal{G}$ acting on $\mathcal{H}$ by isometries.

CONDITIONS 8.1.

1. *There is an action of the Virasoro algebra $\mathfrak{vir}_{\mathbb{K}}$ on ζ, covering the infinitesimal action of* $\mathrm{Diff}^+(S^1)$ *on LX.*

2. *On each component $L_{[g]}X$ of LX, there is a bundle decomposition:*

$$\zeta | L_{[g]}X \cong \widehat{\bigoplus_r} \xi_{[g],r},$$

where the sum is over rational numbers r and each $\xi_{[g],r} \to L_{[g]}X$ is the:

   a. *L_0-eigenbundle for the eigenvalue $-r$,*

   b. *flat bundle associated with a finite-dimensional representation space $W^{\xi}_{g,r}$ of $C_\Pi(g)$.*

3. *For each conjugacy class $[g]$, there exists a natural number $m_{[g]}$ such that in the decomposition, above $r \in (1/m_{[g]})\mathbb{Z}$ and $\xi_{[g],r} = \{0\}$ for all $r \leqslant m_{[g]}$.*

Condition 3 states that ζ is admissible in the sense of Brylinski over each component of LX.

DEFINITION. The character of the admissible bundle ζ is given by characters of the constituent bundles ξ, and:

$$\mathrm{char}_q \xi = \sum_{r \in \mathbb{Q}} \chi_{C_\Pi(g)} W^{\xi}_{\alpha,r} q^r.$$

As always $q = e^{2\pi i \tau}$, and we consider characters as functions on the upper half-plane H_1.

It is possible to give similar conditions for a Hilbert bundle over the double-loop space $L^2X = L(LX)$, acted on by the full diffeomorphism group of the torus $\mathbb{T}^2$. However it suffices in Chap. 8 to consider $\mathrm{Diff}^+(S^1)$ and $\mathfrak{diff}_{\mathbb{K}}$ actions associated with certain fixed point subsets $L^2_{r,s}X$ (defined later) for diagonal actions of the circle diffeomorphism group on subtori in $\mathbb{T}^2$ of rational slope. To make this precise, we use the letter F to denote elements of the double-loop space on which $\mathbb{T}^2$ acts by:

$$((u,v) \cdot F)(z,w) = F(zu, wv).$$

For each pair of coprime integers $r,s > 0$ consider the subtorus:

$$\mathbb{T}_{r,s} = \{(u,v) \in \mathbb{T}^2 : u^s = v^r\} \subseteq \mathbb{T}^2,$$

which we can parametrize by the isomorphism (r and s are coprime!):

$$\begin{aligned} \mathbb{T} &\cong \mathbb{T}_{r,s} \\ t &\mapsto (t^r, t^s). \end{aligned}$$

This group has the fixed point set:

$$L^2_{r,s}X = (L^2X)^{\mathbb{T}_{r,s}} = \{F : F(t^r z, t^s w) = F(z,w) \forall t,z,w \in \mathbb{T}\}.$$

Such an F is determined by its values at $z = 1$, and moreover for all w, we have

$$F(1, \zeta_r^s w) = F(1,w),$$

with $\zeta_r = \exp(2\pi i/r)$. Hence the function defined on $\mathbb{R}$ by the formula:

$$x \mapsto F(1, e^{2\pi i x})$$

has period $1/r$; we can explicitly identify spaces $L^2_{r,s}X$ and LX by means of the map:

$$F \leftrightarrow (z \mapsto F(1, z^{1/r})).$$

Similarly the function:

$$x \mapsto F(e^{2\pi i x}, 1)$$

has period $1/s$, and there is an alternative identification of $L^2_{r,s}X$ with LX by means of the map $F \leftrightarrow (z \mapsto F(z^{1/s}, 1))$. Since we will eventually set $s = 1$, we choose to work with the first of these maps from now on.

Remark. If we fix a base point (x_0, y_0) on $\mathbb{T}_{r,s}$, there is an orthogonal circle $\mathbb{T}_{r',s'}$ with $rs' - sr' = 1$ that meets the first at the common base point. As this varies on $\mathbb{T}_{r,s}$, orthogonal circles partition the 2-torus.

The connection with centralizers of elements in Π arises as follows: an element $F \in L^2_{r,s}X$ satisfies the functional equation:

$$F(u, v\zeta_r^{-s}) = F(u,v),$$

so that if $s's \equiv -1 \pmod{r}$,

$$F(u, v\zeta_r) = F(u, v(\zeta_r^{-s'})^s) = F(u,v).$$

Similarly $F(u\zeta_s, v) = F(u,v)$, and elements of $L^2_{r,s}X$ can be viewed as Π-equivalence classes of maps $\widetilde{F} : \mathbb{R}^2 \to \widetilde{X}$, for which there exist elements $g, h \in \Pi$ satisfying $\widetilde{F}(x+1/s, y) = g\widetilde{F}(x,y)$ and $\widetilde{F}(x, y+1/r) = h\widetilde{F}(x,y)$. Elements g and h commute, since factoring by the action of Π, we obtain a map into X from the torus obtained by identifying opposite sides of the rectangle $[1/s, 0] \times [0, 1/r]$. Note: The holonomy along each *unit* length parallel to the two axes is g^s and h^r, respectively.

DEFINITION. The Hilbert bundle $\zeta \to L^2X$ is *admissible* if for each pair of coprime integers (r,s), the bundle $\zeta_{r,s}$ is admissible.

EXAMPLE 8.1. The construction we now give is modeled on that of [29, pp. 463–467], but it allows a nontrivial fundamental group. Let M be a connected, oriented, smooth manifold and $\xi \to M$ a d-dimensional bundle with characteristic classes w_1, w_2 and $p_1/2$ all equal to zero. Suppose further that ξ is given a Riemannian structure and compatible connection ∇. As before we decompose the loop space LM as:

$$LM = \coprod_{[g]} L_g\widetilde{M}/C_\Pi(g).$$

Let $q : P \to M$ denote the given spin bundle of ξ with structure group Spin_d containing the holonomy group $\mathrm{Hol}(M)$. Let H_0 be the identity component of the closure H of $\mathrm{Hol}(M)$; there is a surjection of Π onto the finite group H/H_0. The following diagram of groups defines $\widetilde{H}$ as a pullback:

$$\begin{array}{ccc} H_0 & = & H_0 \\ \downarrow & & \downarrow \\ \widetilde{H} & \longrightarrow & \Pi \\ \downarrow & & \downarrow \\ H & \longrightarrow & \overline{\Pi} = H/H_0. \end{array}$$

Images on the right of elements on the left are denoted by $[h]$

If $f \in L_g\widetilde{M}$ the relation $hol(f)H_0 = \overline{g} \in \overline{\Pi}$ defines the holonomy of f as an element of H. Motivated by our earlier definition of L_gM, we can write

$$\widetilde{P}_g = \{F : \mathbb{R} \to P : qF \in L_g\widetilde{M} \text{ and } F(t+1) = hol(qF)F(t)\}.$$

We now have a principal fibration $\widetilde{q} : \widetilde{P}_g \to L_g\widetilde{M}$, and the disjoint union over conjugacy classes $[g]$ admits an action by the preceding group $\widetilde{H}$. As expected $h \in \widetilde{H}$ maps $\widetilde{P}_g$ to itself if and only if $[h] \in C_\Pi(g)$.

Now suppose we have a projective representation V of the affine Lie algebra $\widehat{\mathfrak{spin}}_d$ that integrates to a completion $\widehat{V}$ carrying a projective representation of the loop group $L\widehat{\mathrm{Spin}_d}$ of level ℓ say. We can characterize a bundle ξ_V over $\widetilde{P}$ by specifying its sections over a typical open set U to be equivariant maps from U into V. Here equivariance is with respect to the semidirect product of the extended loop group fibering $\widetilde{P}$ and the group $\mathrm{Diff}^+_{\mathbb{R}}$ of orientation-preserving periodic diffeomorphisms of the line:

$$\mathrm{Diff}^+_{\mathbb{R}} = \{\varphi : \mathbb{R} \to \mathbb{R} : \varphi \text{ invertible}, \varphi(0) = 0, \varphi(t+1) = \varphi(t) \forall t \in \mathbb{R}\}.$$

Since the Virasoro algebra is dense in the Lie algebra of a central extension of $\mathrm{Diff}^+_{\mathbb{R}}$, the preceding bundle is $\mathfrak{vir}$-equivariant. Furthermore the L_0-eigenspaces V_r provide the natural grading on V; these are dense in $\widehat{V}$, and as a representation V_r is associated with a finite-dimensional Spin_d-bundle. This becomes flat if we replace V by the invariant subspace V^{H_0}, since its structural group is now the finite group H/H_0. If the original bundle is flat, we constructed an admissible bundle as defined at the beginning of Sec. 8.4, the bundle over the component $L_{[g]}(B(H/H_0))$ is associated with a projective representation of $C_{H/H_0}(g)$. In the special case when H/H_0 is a finite nonabelian simple group, for cohomological reasons the distinction between projective and honest representations disappears. Note: We are working with a *fixed* central extension of H/H_0 by $\mathbb{T}$.

8.5. Constructing Moonshine-like Virasoro-Equivariant Vector Bundles

In this section we establish a framework into which elliptic systems for M_{24} described in Sec. 8.3 naturally fit. To avoid technical problems, we suppose the finite group G is such that

$$H^2(BG, \mathbb{F}_2) = 0, \qquad H^3(BG, \mathbb{Z}) = H^4(BG, \mathbb{Z}) = 0.$$

The first condition ensures that an orientable representation of G satisfies the Spin condition; the other two force projective representations to be genuine representations and some bundles that arise to be admissible. For at least one simple group of interest, namely, Mathieu group M_{23}, these are satisfied. However once a more refined theory is developed, these conditions will probably be modified. Note: Several of our results are already valid only for groups of odd order or after localization away from 2. With this in mind we also note that:

$$H^4(BM_{24}, \mathbb{Z})_{(\text{odd})} = 0,$$

but

$$H^4(BM_{24}, \mathbb{F}_2) \neq 0.$$

The first result is discussed in Chap. 5; the nontriviality of the second group is seen from a computation of the fourth Stiefel–Whitney class of a real representation.

In the discussion that follows, we place the construction of η- and θ/η-examples for M_{24} in a more general setting. As already remarked, the second of these only satisfies the genus zero condition at odd primes; understanding its behavior at the prime 2 needs more work.

Let V be a finite-dimensional real representation of G with rational (and hence integral) character. If $\zeta_V \to BG$ denotes the associated flat bundle, for any loop $f : \mathbb{T} \to BG$ (or double loop $F : \mathbb{T}^2 \to BG$) we have the space of smooth sections $\Gamma(\mathbb{T}, f^*\zeta_V)$ (or $\Gamma(\mathbb{T}^2, F^*\zeta_V)$) of the pullback bundle. A specific section is given by a G-equivalence class of maps defined on $\mathbb{R}$ of the form:

$$t \mapsto (\widetilde{f}(t), v(t))$$

where $v : \mathbb{R} \to V$ is smooth, $\widetilde{f}$ is the lift of f to EG, and both $\widetilde{f}$ and v are periodic with respect to the holonomy of f contained in the class $[g]$. Since g has finite order $|g|$, $v(t+|g|) = v(t)$, and the function v has a Fourier expansion:

$$v(t) = \sum_{k \in \mathbb{Z}} v_{k/|g|} e^{2\pi i k t/|g|},$$

for $v_{k/|g|} \in V \underset{\mathbb{R}}{\otimes} \mathbb{C}$ satisfying $g v_{k/|g|} = \zeta^k_{|g|} v_{k/|g|}$: Therefore:

$$v_{k/|g|} \in V^g_{k/|g|} = \{u \in V : gu = \xi^k_{|g|} u\}.$$

For a double loop $(x,y) \mapsto (\widetilde{F}(x,y), v(x,y))$ with v having a Fourier expansion:

$$v(x,y) = \sum_{k,\ell \in \mathbb{Z}} v_{k/|g|,\ell/|h|} \exp\left(2\pi i \left(\frac{kx}{|g|} + \frac{ly}{|h|}\right)\right),$$

the coefficients satisfy

$$g v_{k/|g|,\ell/|h|} = \zeta^k_{|g|} v_{k/|g|,\ell/|h|} \qquad h v_{k/|g|,\ell/|h|} = \zeta^\ell_{|h|} v_{k/|g|,\ell/|h|}.$$

Restricting to the fixed point set $L^2_{r,s}BG$ and recalling that the restricted double loop F has a period $1/s$ in the first factor and $1/r$ in the second, we have a Fourier expansion:

$$v(x,y) = \sum_{k,\ell \in \mathbb{Z}} v_{ks/|g|,\ell r/|h|} e^{2\pi i (ksx/|g| + \ell r y/|h|)}.$$

Coefficients in $V \underset{\mathbb{R}}{\otimes} \mathbb{C}$ satisfy

$$g v_{ks/|g|,lr/|h|} = \zeta^{ks}_{|g|} v_{ks/|g|,lr/|h|} \text{ and } h v_{ks/|g|,lr/|h|} = \zeta^{lr}_{|h|} v_{ks/|g|,lr/|h|}.$$

If $\widetilde{F}$ is a lift of F having holonomy g along the interval of length $1/s$ in the x-direction and h along an interval of length $1/r$ in the y-direction, then the function v satisfies $v(x+1/s,y) = g(x,y)$ and $v(x,y+1/r) = h(x,y)$. This periodicity is equivalent to:

$$gv_{k,l} = \zeta_s^k v_{k,l} \qquad hv_{k,l} = \zeta_r^l v_{k,l}.$$

Note: We recover periods equal to 1 for the original section of the double-loop space at the price of replacing g and h by g^s and h^r, respectively.

The space $L^2_{r,s}BG$ is replaced with the standard single-loop space LBG by the correspondence $F \leftrightarrow (z \mapsto F(1,z^{1/r}))$; over a loop $f : \mathbb{T} \to BG$, the fiber of the corresponding pullback bundle consists of maps:

$$\mathbb{R} \to EG \underset{G}{\times} V$$

$$t \mapsto \left[\widetilde{F}(0,t/r), \sum_{j,k} v_{j,k} e^{2\pi i k t/r} \right]_G$$

where $\widetilde{F}$ lifts $F : \mathbb{R}^2 \to BG$. Summing over the index j in the second component, $v_k = \sum_j v_{j,k}$. Under the assumption that a lift of f has holonomy g^* this implies $g^* v_k = \zeta_r^k v_k$.

Including the action of $\mathrm{Diff}^+(S^1)$ and hence of the infinitesimal generator L_0, we see that:

$$L_0\left(\left[\widetilde{F}(x,y), \sum_{j,k\in\mathbb{Z}} v_{j,k} e^{2\pi i (jrx+ksy)/rs}\right]\right)$$

$$= \left[\widetilde{F}(x,y), \sum_{j,k\in\mathbb{Z}} \frac{-(jr+ks)}{rs} v_{j,k} e^{2\pi i (jrx+ksy)/rs}\right].$$

Briefly stated eigenspaces of L_0 correspond to rational numbers of the form $(jr+ks)/rs$, although we said nothing about lower bounds, i.e., about admissibility. Assuming further that $s = 1$, the preceding calculation implies that sections over $f \in LBG$, associated to $\widetilde{f}$ with holonomy $g \in G$ take the form:

$$\sum_{k\in\mathbb{Z}} v_{k/|g|} z^{kr} \text{ with } v_{k/|g|} \in V^g_{k/r} = \{v \in V : gv = \zeta_r^k v\}.$$

The subspace $V^g_{k/r}$ carries a representation of the centralizer $C_G(g)$; therefore as in Sec. 8.3 the space of sections can be viewed as a bundle over LBG. On the component of loops with holonomy conjugate to g, this is restricted to a completed sum of flat bundles, each of which is an eigenbundle for L_0.

With rational grading k/r, we now have the completed sum:

$$V_q^{[g]} = \widehat{\bigoplus}_{k\in\mathbb{Z}} V_{k/r}^{[g]} q^{k/r} \in K^0(BC_G(g))[[q, q^{-1}]].$$

We can likewise form the so-called Fock space:

$$\prod_{\substack{0\leqslant k\leqslant r-1\\ 0\leqslant l\in\mathbb{Z}}} S(V_{k/r}^{[g]} q^{(lr+k)/r});$$

compare the construction of the inverse representation to that with character $\eta_g(\tau)$ in the previous section.

We come closer to Mason's example of an elliptic system if we base our construction on a lattice L, with an even-valued quadratic form on which the finite group G acts. In the general setting, we borrow notation and terminology from [46] and quote the results where necessary. As in the case of the permutation representation of M_{24}, we set $L_{2n} = \{l \in L : Q(l) = 2n\}$ and write L for the disjoint union of the sets L_{2n}.

Consider a central extension:

$$1 \longrightarrow \langle\kappa\rangle \longrightarrow \widehat{L} \xrightarrow{\pi} L \longrightarrow 1$$

where κ has order 2, and the 2-cocycle $\varepsilon_0 : L\times L \to \mathbb{Z}/2$ satisfies $\varepsilon_0(l_1,l_2) - \varepsilon_0(l_2,l_1) = c_0(l_1,l_2)$, with $c_0(l_1,l_2) = \langle l_1,l_2\rangle \pmod 2$. Here $\langle,\rangle$ denotes the inner product on L associated with the quadratic form. Choose a section $s : L \to \widehat{L}$, which sends $O \in L$ to the identity element in $\widehat{L}$. Use the grading on L to define one on $\widehat{L}$ (so that the cardinality of $\widehat{L}_{2n}$ equals twice that of L_{2n}). For each $n \geqslant 0$, we form the free $\mathbb{K}$-module on elements of L_{2n} ($\widehat{L}_{2n}$), and combining these we obtain graded group rings $\mathbb{K}[L]_*$ ($\mathbb{K}[\widehat{L}]_*$). If N is any graded $\mathbb{K}[\langle\kappa\rangle]$-module, we can form the left $\mathbb{K}[\widehat{L}]_*$-module $\mathbb{K}[\widehat{L}]_* \otimes_{\mathbb{K}[\langle\kappa\rangle]} N_*$ with its grading as a tensor product. For example we can give $\mathbb{K}$ the module structure $\kappa\cdot x = -x$ (denoted $\mathbb{K}^-$) with grading 0 and write

$$\mathbb{K}\{L\} = \mathbb{K}[\widehat{L}]_* \underset{\mathbb{K}[\langle\kappa\rangle]}{\otimes} \mathbb{K}^- \cong \mathbb{K}[L]_*$$
$$s(l)\otimes 1 \leftrightarrow l.$$

Note: $\kappa(s(l)\otimes 1) = -(s(l)\otimes 1)$. The action of G on L now induces an action on $\mathbb{K}\{L\}$.

Let $\mathfrak{h}\underset{\mathbb{Z}}{\otimes} L$ and define $S(\widehat{\mathfrak{h}}_{\mathbb{Z}}^-)$ to be the suitably graded symmetric algebra. Finally form the graded $\mathbb{K}[G]$-module:

$$V_{L_*} = \mathbb{K}\{L\}_* \underset{\mathbb{K}}{\otimes} S(\widehat{\mathfrak{h}}_{\mathbb{Z}}^-).$$

PROPOSITION 8.2. *The module V_{L_*} admits an action by the Virasoro algebra $\mathfrak{vir}_{\mathbb{K}}$.*

For a proof, see [46], particularly chaps. 7 and 8. The details would take us too far afield from the matter of this book, but it is important to show that Proposition 8.2 brings the elliptic objects constructed *ad hoc* for M_{24} into the framework of $TQFT$.

We are now ready to combine two constructions. The original lattice L determines a representation space V to which we can associate a flat bundle $EG \underset{G}{\times} V \to BG$. Pulling back along a loop $f : \mathbb{T} \to BT$ and taking sections, i.e., G-equivalence classes of maps of the form:

$$t \mapsto (\widetilde{f}(t), v(t)),$$

we obtain a family of fibrations with fibers $(L \underset{\mathbb{Z}}{\otimes} \mathbb{R})^g$ and structural groups $C_G(g)$ as $[g]$ runs through all possible conjugacy classes. We can also form the Fock bundles with fibers $S(\widehat{\mathfrak{h}}^-)$ and the tensor product over $\mathbb{R}$ with $\mathbb{C}\{L\}$. On each fiber this bundle agrees with the module $V_{L_*^g}$, as previously constructed, but with group G replaced by $C_G(g)$. The result of our labors is a Virasoro-equivariant bundle over LBG with character:

$$q^{\operatorname{rank} L^g/24}\theta'_{L^g}/\eta_g(L \otimes \mathbb{R})^g.$$

The normalization factor $q^{\operatorname{rank} L^g/24}$ appears so that grading starts at 0 and the q-series has the required modularity properties. Evaluating this character at the identity element and neglecting normalization, we have Proposition 8.3.

PROPOSITION 8.3.

$$\theta'_{L^g}/\eta_g(L \oplus \mathbb{R})^g(1) = \left(\sum_{n \geqslant 0} |L_{2n}| q^n\right)\left(q^{1/24}\prod_{n \geqslant 1}(1-q^n)\right)^{-\operatorname{rank} L^g}$$

8.6. Notes

The origin of moonshine constructions is the observation in [36] that coefficients of the modular function:

$$j(\tau) - 744 = q^{-1} + 196884q + \cdots$$

are closely related to degrees of irreducible representations of the monster simple group $\mathbb{M}$. This leads one to ask whether it is possible to associate to each conjugacy class $[g]$ in $\mathbb{M}$ a modular form $j_{[g]}(\tau)$, such that:

***M*1:** The family of functions $j_{[g]}(\tau)$ form the character of a Thompson series, and

***M*2:** For each class $[g]$ there is a discrete group $\Gamma_g \subseteq SL_2(\mathbb{Z})$ having compactified orbit space $(\Gamma_g \backslash SL_2(\mathbb{Z}))^-$ of genus one and function field generated by $j_{[g]}$.

As a spin-off of this work, similar conjectures can be made for certain other groups, notably those arising as centralizers in $\mathbb{M}$. In a truly remarkable short appendix to the survey [83], Norton proposed combining the then existing phenomena into a single 2-variable moonshine, and formulated the equivariant modularity condition that now plays a key role in Devoto's coefficient ring $\mathcal{E}ll^*_G$.

The numerology associated with finite simple groups was developed independently of elliptic cohomology, and it was only as a consequence of the work of Hopkins, Kuhn, and Ravenel [56, 57] as interpreted by Segal [101] that it was realized Thompson series should be regarded as preferred elements in $Ell^*(BG)$. Segal's interest in conformal field theory was significant at this point — in particular his imaginative leap from vector bundles over X as functors from a category of points and paths in X to finite-dimensional vector spaces to graded vector bundles over LX, a path interpreted as the image of a Riemann surface. In our treatment we follow the survey by Atiyah [14], modified by a reading of both [101, 94]. Two additional comments are worth making: Firstly the use of the double-loop space to obtain 2-variable Virasoro bundles points to the existence of $K3$ or string cohomology, where abelian surfaces play the role of elliptic curves (see Chap. 10). Secondly although we concentrate on the moonshine approach, the reader should be aware of [29], to which we refer only in passing.

Viewed from elliptic cohomology, the construction of the moonshine module $V^\natural = V^+ \oplus V'^-$ in [46] appears from the bottom up. The importance of the modular forms η_g is that for the group M_{24}, these arise naturally as generators of one-dimensional eigenspaces for the Hecke algebra in its guise as cohomology operations in Ell^*. Combined with the character of $\mathbb{C}\{L\}$, where L is either the natural permutation module for M_{24} or the Leech lattice for Co_0, we obtain a candidate for a moonshine module for these two groups. To construct the summand V^+ of $V^\natural$ we start from this module, that is one from the class constructed in Sec. 8.4. For the summand V'^- we start with a representation space for the subgroup $O_2(C_{\mathbb{M}}(t))$ where t is a so-called Type 2 involution in the monster, and $O_2(C)$ is an extra-special subgroup of order $2^{(2.12+1)}$ with center equal to $\langle t \rangle$. Again we see that the pattern for odd primes is clear and already laid down for M_{24}. To help explain what happens at the prime 2 in this construction and elsewhere, we must define elliptic cohomology over the ring $\mathbb{Z}$ rather than over $\mathbb{Z}[1/2]$. We start to do this in Chap. 9.

9

Variants of Elliptic Cohomology

Occasionally in earlier chapters we refer to elliptic cohomology of level 1, that is, to a theory where the coefficient ring consists of modular forms invariant with respect to the group $SL_2(\mathbb{Z})$ rather than its subgroup $\Gamma_0(2)$. Because the coefficients are easily described and well-understood, this level 1 theory is no more difficult than the theories with coefficients $\mathbb{Z}[1/2][\delta, \varepsilon, \varepsilon^{-1}]$ and $\mathbb{Z}[1/2][\delta, \varepsilon, \Delta^{-1}]$. Using more complex results about modular forms, it is also possible to define level N theories, and we do this in Sec. 9.1.

The main part of Chap. 9 is devoted to extending the definition of $Ell^*(X)$ to the prime 2, that is, to showing that there exists a theory defined over $\mathbb{Z}$ as opposed to $\mathbb{Z}[1/2]$. We actually define a homology rather than a cohomology theory, and the motivation comes from spin bordism and the observation that inverting $\mathbb{H}P^2$ is rather like inverting ε in the preceding coefficient ring. Once we formalize this, the cases $p =$ odd and $p = 2$ must be discussed separately. In the former it is necessary to construct specific generators for spin bordism groups in terms of three-dimensional quaternionic bundles over base manifolds having a spin structure. The argument is technically nontrivial, but it uses standard techniques from algebraic topology. The latter is much more interesting, since at the start there is less sense of what we are seeking. After borrowing a result from manifolds of positive scalar curvature, we modify the cofiber of a map of spectra

$$M\mathrm{Spin}_\wedge \Sigma^8 B(PSp_3) \to M\mathrm{Spin}.$$

The group PSp_3 appears as the isometry group of $\mathbb{H}P^2$ with its usual metric. In Secs. 9.2–9.5 we outline the main steps of the proof.

The final part of Chap. 9 introduces the work of Hopkins [55] on the spectrum eo_2. This starts from the observation that E. Witten's version of the elliptic genus takes only values in a ring of forms that are strictly modular if the domain is restricted to manifolds where $(1/2)p_1$ as well as w_1 and w_2 vanish. We already met this condition as one that ensures the existence of a spin structure on the infinite-dimensional manifold LM. One technical problem is that a single line bundle does not admit the necessary geometric structure, forcing us to replace the characteristic series $Q(x)$ for a genus with the notion of a cubical structure. Assuming this, it is possible compatibly to orient a family of elliptic spectra, so that the natural target is the limit spectrum eo_2. As [55] comments: "This cohomology theory can be used to account for nearly everything that is known about the stable homotopy groups of spheres in dimensions less than 60."

9.1. Elliptic Genera of Level N

In constructing a level 1 theory, our model is the version of elliptic cohomology with coefficients obtained by inverting the discriminant Δ rather than ε. Thus the level 1 coefficient ring is

$$\mathbb{Z}\left[\frac{1}{6}\right][g_2, g_3, \Delta^{-1}]$$

where the degree of $g_i = 4i (i = 2, 3)$ and $\Delta = g_2^3 - 27g_3^2$.

Using the universality of the formal group law for complex (as opposed to oriented real) cobordism, the group law associated with the Weierstrass curve $y^2 = 4x^3 - g_2x - g_3$ requires a homomorphism:

$$\varphi^1 : \Omega_U^* \to Ell_1^* = \mathbb{Z}\left[\frac{1}{6}\right][g_2, g_3, \Delta^{-1}],$$

with which we define the required Ω_U^*-structure on $Ell_1^* = Ell_1^*$ (point). Landweber conditions for a cohomology theory are satisfied [51]. As before, we use the fact that the formal group law on any elliptic curve has height equal to at most 2. The summand Ell_1^{2n} consists of modular forms of weight n invariant under the action of the whole modular group $SL_2(\mathbb{Z})$. These are meromorphic at the single cusp $i\infty$, and their Fourier coefficients are in $\mathbb{Z}[1/6]$. Here we use the identifications:

$$g_2 \leftrightarrow \frac{1}{12}E_4 \qquad g_3 \leftrightarrow -\frac{1}{216}E_6,$$

where E_{2n} stands for the Eisenstein series:

$$E_{2n}(q) = 1 - \frac{4n}{B_{2n}} + \sum_{k\geqslant 1}\left(\sum_{d|n} d^{2k-1}\right)q^n,$$

and $B_2, B_4, \ldots$ are Bernoulli numbers.

For theories of higher weight N, the main step is again to describe the necessary structural map from complex cobordism. As usual let $\tau \in H_1$ and L be the lattice $2\pi i(\mathbb{Z}\tau + \mathbb{Z})$. We seek a complex function $h(z)$ that is elliptic with respect to L and on the orbit space $\mathbb{C}/L$ has a zero of order N at the origin O and a pole of order N at some point P. The condition $N \cdot O - N \cdot P \equiv O \pmod{L}$ implies that P must be a nonzero N-division point of $\mathbb{C}/L$, i.e., $P = 2\pi i\,[(k\tau + l)/N] \neq 0$ for $k, l \in \mathbb{Z}$.

Normalize h by $h(z) = z^N + \cdots$ and extract an Nth root $f(z) = \sqrt[N]{h(z)} = z + \cdots$, which is elliptic with respect to a sublattice $\overline{L}$ of index N in L. A candidate for the characteristic function associated with a genus is given by:

$$Q(z) = \frac{z}{f(z)},$$

a power series whose coefficients b_r as functions of the lattice variable τ are modular forms of weight r. If P is a primitive N-division point, the invariance subgroup equals $\Gamma_1(N)$, consisting of matrices congruent to $\left(\begin{smallmatrix} 1 & b \\ 0 & 1 \end{smallmatrix}\right) \pmod{N}$. Now formally factorize the total Chern class of a stably complex manifold M^{2d} as $c.(M) = 1 + c_1(M) + \cdots + c_d(M) = (1 + x_1) \cdots (1 + x_d)$ and define

$$\varphi_{N,P}(M) = \left(\prod_{j=1}^{d} x_j / f(x_j) \right) [M].$$

This is an expression of weight d in coefficients of $Q(z)$, hence a modular form of weight d with invariance subgroup $\Gamma_1(N)$. Here we assumed that $P = 2\pi i\ell/N$ with $0 < \ell < N$ and $(\ell, N) = 1$. If $(\ell, N) = n \neq 1$ we obtain a larger invariance subgroup $\Gamma_1(N/n) \supset \Gamma_1(N)$.

REMARK. One reason why $N = 2$ is special is that in this case, we can use an expansion in terms of x_i^2 and so work with Pontrjagin rather than with Chern classes.

To be more precise about the ring where this genus takes its values, we write

$$Ell\left[\frac{1}{N}\right]^* = \mathbb{Z}\left[\frac{1}{6N}\right][E_4, E_6, \Delta^{-1}],$$

where in accordance with the identifications of g_i with E_{2i} $(i = 2, 3)$:

$$\Delta = \frac{1}{1728}(E_4^3 - E_6^2).$$

Note: We inverted N as well as 6, since our forms are to be meromorphic at each cusp, with Fourier expansions in $\mathbb{Z}[1/6N, \zeta_N]((q^{1/N}))$. The relation between $Ell[1/N]^*$ and the ring $Ell^*_{\Gamma_1(N)}$ just described is given in terms of Galois theory. Write $\Gamma(N)$ for matrices congruent to the unit matrix (modulo N). Then we have Proposition 9.1.

PROPOSITION 9.1. *The ring $Ell^*_{\Gamma(N)}$ is a finitely generated Galois extension of $Ell[1/N]^*$ with Galois group $SL_2(\mathbb{Z})/\Gamma(N) \cong SL_2(\mathbb{Z}/N)$. The subgroup $\Gamma_1(N)/\Gamma(N)$ has fixed ring $Ell^*_{\Gamma_1(N)}$.*

Proof: See [18, Theorems 1.1 and 1.2]. ■

There are now two ways of defining level N elliptic cohomology, using either the structural map $\varphi_{N,P}$ just described, or the composition:

$$\Omega^*_U \underset{\varphi^1}{\longrightarrow} Ell^* \underset{i}{\hookrightarrow} Ell^*_{\Gamma_1(N)}.$$

Baker shows that these genera are associated with strictly isomorphic formal group laws; that is enough to check Landweber's exactness conditions for the composition $i\varphi^1$.

PROPOSITION 9.2. *The tensor product $\Omega^*_U(X) \underset{i\varphi^1}{\otimes} Ell^*_{\Gamma_1(N)}$ is a multiplicative complex-oriented cohomology theory on finite CW-complexes X.*

Proof: For each prime $p > 3$ for which $p \nmid N$, the sequence p, v_1, v_2 in Ell^* remains regular in the extension ring $Ell^*_{\Gamma_1(N)}$. Modulo $\langle p, v_1 \rangle$ the class of v_2 remains a unit, so it is enough to look at the pair (p_1, v_1). The ring $Ell^* / \langle p \rangle$ is a principal ideal domain (graded); if the class of v_1 were annihilated by the class of some element u, then the constant term in the minimal polynomial of u would also annihilate v_1 (mod p). This implies that u equals 0 (mod p), so we are done. ■

This composition approach also applies to the principal congruence subgroup $\Gamma(N)$. In fact using more delicate techniques [29] obtained level N theories from ordinary elliptic cohomology [associated with $\Gamma_0(2)$] by exploiting the fact that the required finitely generated extension rings of modular forms are *faithfully flat*. The advantage over Baker's approach is that Fourier coefficients belong to $\mathbb{Z}[1/2, \zeta_N]$, so we need not invert N. However to prove first that the ring of modular forms is finitely generated over this smaller algebra and secondly that the faithfully flat condition holds [29, Theorem 3.5 and Proposition 3.6], he must use delicate results about moduli schemes of elliptic curves with level structures. Allowing denominators that divide N thus leads to a much more elementary argument.

9.2. Projective Plane Functors

In this section we formulate the definition of level 2 elliptic *homology* in a more general setting. The results are due to [67] and so far as the elliptic case is concerned, these are contained in [67]. Let $\mathbb{K}$ be $\mathbb{R}$, $\mathbb{C}$, $\mathbb{H}$, or $\mathbb{O}$ (the Cayley numbers). As $\mathbb{K}$ varies consider the following bordism theories defined on a suitable category of CW-complexes:

$\mathbb{R}$	$N_n(X)$	unoriented bordism	
$\mathbb{C}$	$\Omega_n^{SO}(X)$	oriented bordism	$w_1 = 0$
$\mathbb{H}$	$\Omega_n^{\mathrm{Spin}}(X)$	spin bordism	$w_1 = w_2 = 0$
$\mathbb{O}$	$\Omega_n^{\langle 8\rangle}(X)$	$\widetilde{\mathrm{spin}}$ bordism	$w_1 = w_2 = \frac{1}{2}p_1 = 0.$

Conditions on characteristic classes in the last example ensure that the classifying map for the tangent bundle τM^n factors through the 7-connected covering space $BO\langle 8\rangle$ of BO. For each division algebra consider the diagram:

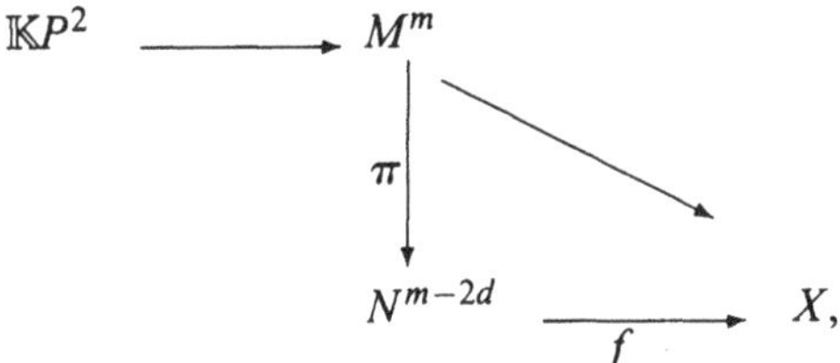

where $d = 1, 2, 4$, or 8, depending on the choice of $\mathbb{K}$, and π is the projection map of a fiber bundle having structural group equal to the isometries of $\mathbb{K}P^2$. Once more depending on the choice of $\mathbb{K}$, this equals PSO_3, PU_3, PSp_3, or F_4.

Definition. Let $T_n^{\mathbb{K}}(X)$ equal the subgroup of $\Omega_n^{\mathbb{K}}(X)$ consisting of bordism classes $[M, f\pi]$ and let $\widetilde{T}_n^{\mathbb{K}}(X)$ equal the subgroup whose elements satisfy the additional assumption that $[N, f] = 0$ in $\Omega_{n-2d}^{\mathbb{K}}(X)$.

Definition. The $\mathbb{K}$-projective plane functors are given by:

$$X \mapsto pE_n^{\mathbb{K}}(X) = \Omega_n^{\mathbb{K}}(X)/T_n^{\mathbb{K}}(X),$$
$$X \mapsto \pi E_n^{\mathbb{K}}(X) = \Omega_n^{\mathbb{K}}(X)/\widetilde{T}_n^{\mathbb{K}}(X).$$

It is conjectured, and in most cases proved, that in all cases pE_n and πE_n are homology theories. What interests us most at present is the π-theory for $\mathbb{K} = \mathbb{H}$, so from now on we drop the affix $\mathbb{H}$. We also allow localization with respect to some suitable multiplicative subset.

The Cartesian product of manifolds induces a multiplication:

$$\pi E_m(X) \times \pi E_n(Y) \to \pi E_{m+n}(X \times Y)$$

and a natural transformation:

$$\Omega_*^{\mathrm{Spin}}(X) \underset{\Omega_*^{\mathrm{Spin}}}{\otimes} \pi E_* \longrightarrow \pi E_*(X). \qquad (\natural)$$

As usual πE_* means πE_* (point) and the preceding transformation is compatible with multiplication. The main result that we wish to prove is Theorem 9.1.

THEOREM 9.1.

1. *$\pi E_*(X) \otimes \mathbb{Z}_{(2)}$ is a multiplicative homology theory.*
2. *The natural transformation ($\natural$) becomes an isomorphism after inverting 2.*
3. *The coefficients $\pi E_* \cong \mathbb{Z}[s,k,h,b]/\langle 2s, s^3, sk, k^2 - 4(b+64h)\rangle$.*

REMARK. The labels of generators for πE_* are borrowed from their natural preimages in Ω_*^{Spin}, the classes of S^1 (with the nontrivial spin structure), the Kummer surface K^4 (unique up to diffeomorphism with signature 16), B^8 (with $\hat{A} = 1$ and signature 0), and $\mathbb{H}P^2$ (with $\hat{A} = 0$ and signature 1). Relations in πE_* follow from those holding between spin-manifolds.

Consider the effect of inverting h in $\pi E_*(X)$, i.e., in defining:

$$\pi E_n(X)[h^{-1}] = \lim_j \pi E_{n+8j}(X),$$

when the limit is taken over the sequence of homomorphisms given by multiplying by h; thus:

$$\pi E_n(X)[h^{-1}] = \bigoplus_{j \in \mathbb{Z}} \Omega_{n+8j}^{\mathrm{Spin}}(X)/\sim,$$

where the equivalence relation $\sim$ is generated by identifying $[N,f]$ with $[M,f\pi]$ for an $\mathbb{H}P^2$-bundle $\pi : M \to N$ and map $f : N \to X$.

THEOREM 9.2. *$\pi E_*(X)[h^{-1}]$ is a homology theory that agrees with the standard level 2 elliptic homology theory $Ell_*(X)$ after inverting 2.*

Proof: Combine Theorem 9.1 with the proof in Chap. 2 that $Ell^*(X)$ is a cohomology theory. Inverting 2 allows us to identify $\pi E[1/2]_*$ with $\mathbb{Z}[1/2][\delta,\varepsilon]$, and the further inversion of h corresponds to inverting ε. Localizing πE_* at 2 reduces the argument to part 1 of Theorem 9.1, since a direct limit of exact sequences is exact and localization at $\{h,h^2,\dots\}$ does not affect the validity of homology axioms. ∎

REMARK. If we replace πE_* by pE_* and invert the generator b we recover real K-theory $KO_*(X)$. This result recalls [34] to which we refer in Chap. 2.

9.3. Atiyah Invariant and the Ochanine Genus

Following [91] we propose to define a genus of elliptic type that simultaneously generalizes a variant of the $\hat{A}$-genus, taking account of the prime 2 and the universal elliptic genus, as discussed in Chaps. 1 and 2. For the first invariant, we recall that an oriented cohomology theory admits a Gysin homomorphism:

$$\pi_!^M : h^*(M) \to h^{*-m}(M)$$

associated with the constant map $\pi^M : M^m \to$ point for every oriented manifold in the theory. With $h^* = KO^*$, a spin manifold is oriented, so we can define

$$\alpha(M) = \pi_!^M(1),$$

where 1 refers to the class of the trivial real-line bundle over M. We recover the $\hat{A}$-genus by composing α with the Pontrjagin character:

$$ph : KO^{-m}(\text{point}) \to \mathbb{Q},$$

which evaluates the Chern character of a complexified bundle on the fundamental class of S^m. [Recall that $KO^{-m}(\text{point}) = \widetilde{KO}(S^m)$ and $ch(\xi) = \dim(\xi) + \sum_j s_j(\xi)/j!$, where s_j is the jth Newton polynomial in the Chern classes.] The advantage of α over $\hat{A}$ is that it allows us to handle the summands $\mathbb{Z}/2$ in KO^* (point).

Consider the diagram:

$$\begin{array}{ccc} & \overset{---\beta--->}{} & \\ \Omega_*^{\text{Spin}} & \xrightarrow[\alpha]{} KO_*(pt) \hookrightarrow & KO_*(pt)[[q]] \\ \varphi_{Ell}\downarrow & & \downarrow ph \\ \mathbb{Z}\left[\tfrac{1}{2}\right][\delta,\varepsilon] & \xrightarrow[i]{} & \mathbb{Q}[[q]] \end{array}$$

where ph is the extension of the Pontrjagin character to a power series, i is the natural embedding that sends δ and ε to their Fourier expansions, and φ_{Ell} is the genus characterized by strong multiplicativity, and $\varphi_{Ell}(\mathbb{C}P^2) = \delta, \varphi_{Ell}(\mathbb{H}P^2) = \varepsilon$. We propose to extend α to a genus β, not localized away from 2, taking values in modular forms over $KO_*(pt)$. We first write

$$W_q(\xi) = \sum_{j\geqslant 0} W^i(\xi)q^i = \bigotimes_{\ell\geqslant 1}(\wedge_{-q^{2\ell-1}}(\xi)\bigotimes S_{q^{2\ell}}(\xi))$$

for any real bundle ξ. Note: W_q is exponential in the sense that:

$$W_q(\xi_1 \oplus \xi_2) = W_q(\xi_1) \otimes W_q(\xi_2);$$

hence it extends to virtual bundles and defines a homomorphism $W_q : KO(X) \to KO(X)[[q]]$.

Definition. The Ochanine genus $\beta(M)$ of an m-dimensional spin manifold is given by:

$$\beta(M) = \sum_{j\geqslant 0} \beta^j(M)q^j = \sum_{j\geqslant 0} \pi_!^M(W^j(\tau M - m))q^j \in KO_m(pt)[[q]].$$

Note: $W^0(\xi)$ always equals the trivial real-line bundle; hence $\beta^0(M) = \pi_!^M(1) = \alpha(M)$. Then we produced a genuine extension of α which from its definition is multiplicative on Cartesian products. Actually more is true; G. B. Segal's proof of the equivalence of rigidity and strong multiplicativity [101, Sec. 3] can be adapted to prove Proposition 9.3.

Proposition 9.3. *Let $\pi : M \to N$ be a PSp_3-bundle over the spin manifold M^m with fiber $\mathbb{H}P^2$. Then $\beta(M) = \beta(N)\beta(\mathbb{H}P^2)$.*

Note: Besides Segal's argument and an *ad hoc* check of rigidity [25] in this case, we need the fact that the complexification map:

$$\operatorname{coker}(KO^{-8}(pt) \to KO_G^{-8}(pt)) \to \operatorname{coker}(K^{-8}(pt) \to K_G^{-8}(pt))$$

is injective. This is clear when filling in the various terms. It is presumably also be possible to construct a proof along the lines of the one given in Chap. 1 for the strong multiplicativity of φ_{Ell}.

9.4. Kernel and Images of the Ochanine Genus

We warm up by considering the restriction to degree zero $\alpha : \Omega_*^{\mathrm{Spin}} \to KO_*(pt)$. The structure of $KO_*(pt)$ is known by Bott periodicity:

$$KO_*(pt) = \mathbb{Z}[\eta, \omega, \mu, \mu^{-1}]/\langle 2\eta, \eta^3, \eta\omega, \omega^2 - 4\mu\rangle,$$

where η, ω, and μ are the images under α of spin generators S^1, K, and B respectively. We already noted that $\hat{A}(\mathbb{H}P^2)$ vanishing implies $\alpha(\mathbb{H}P^2)$ vanishing.

PROPOSITION 9.4. $\ker(\alpha) = T_n(pt)$

Sketch of proof: The vanishing of $\alpha(\mathbb{H}P^2)$ can also be seen as resulting from the existence of a metric of positive scalar curvature on $\mathbb{H}P^2$ (the standard one). By direct construction the same is true for total spaces of $\mathbb{H}P^2$-bundles, i.e., α vanishes on $T_n(pt)$. The proof of the converse at the prime 2 is contained in [104]; at odd primes we appeal to the characterization of generators for $\Omega_*^{\mathrm{Spin}}[1/2](pt)$ in terms of characteristic numbers. It is then necessary to construct specific generators by projectivizing the appropriate three-dimensional quaternionic bundles over $(4n-8)$-spin manifolds for $n \geqslant 2$. These turn out to be

$$(\gamma_1 \times \gamma_2) \oplus \mathbb{H} \longrightarrow \mathbb{H}P^r \times \mathbb{H}P^{n-r-2}$$

for $0 \leqslant r \leqslant n-2$, where γ_1 and γ_2 are canonical quaternionic line bundles over the factors and $\mathbb{H}$ is trivial. Choose the orientation on quaternionic projective space, so that $\langle y^k, [\mathbb{H}P^k] \rangle = 1$, where y pulls back to the square of the generator given by the complex structure on $S^2 \subset \mathbb{C}P^{2k+1}$. We leave the determination of characteristic numbers of associated $\mathbb{H}P^2$-bundles as an exercise. Also check that these fulfill the conditions needed for a complete family of generators in spin-bordism. See [67, Proposition 4.2 and Lemma 4.3]. ■

Consider now the full Ochanine genus β, which is multiplicative for $\mathbb{H}P^2$-bundles by Proposition 9.3. Hence by definition β vanishes on the subgroup $\widetilde{T}_n(pt)$.

PROPOSITION 9.5.

1. $\ker \beta = \widetilde{T}_n(pt)$.
2. *The image of β is generated by images of S^1, K, B, and $\mathbb{H}P^2$, satisfying no other relations than those from spin bordism.*

Proof: We outline two methods of proof, the first of which uses Proposition 9.4

A: The relations already appeared in the statement of Theorem 9.1. Given that $\Omega_1^{\mathrm{Spin}} = \mathbb{Z}/2$ and $\Omega_3^{\mathrm{Spin}} = \Omega_5^{\mathrm{Spin}} = 0$, the only nontrivial relation is $4(b + 64h) = k^2$. This follows from calculating the $\hat{A}$-genus and signature for the square of the Kummer surface — 2^2 and 16^2, respectively. We already know that $\hat{A}(\mathbb{H}P^2) = 0$ [$\hat{A}(B) = 1$] and $L(\mathbb{H}P^2) = 1$ [$L(B) = 0$]. To show that applying β introduces no further relations, we only need the values of α on S^1, K, and B and the value of φ_{Ell} on $\mathbb{H}P^2$. The fact that β is a structure preserving map shows that there can be no more $\mathbb{Z}$- or $\mathbb{Z}/2$-relations. It remains to prove that Ω_n^{Spin} is the sum of a subgroup coming from these same four generators and $\widetilde{T}_n(pt)$. We proceed by induction: There is nothing to prove for $n \leqslant 9$; assume that the claim is true for $n < 8k$ and $[M] \in \Omega_{8k+r}^{\text{Spin}}$ for $0 \leqslant r < 8$. If $r = 0, 1, 2$ ($r = 4$), we can subtract a multiple of $B^k S^{1r}$ ($B^k K$) to achieve $\alpha(M) = 0$. Hence by Theorem 9.4, M is bordant to the total space of an $\mathbb{H}P^2$-bundle over N, implying that $[M] \equiv [N] \times [\mathbb{H}P^2]$ (mod $\widetilde{T}_*(pt)$). This completes the inductive step.

B: It is possible to side-step the appeal to Proposition 9.4, and hence to S. Stolz's results on metrics of positive scalar curvature by noting that in dimensions not congruent to 1 or 2 modulo 8, modular form arguments already used for φ_{Ell} apply to β. If $KO^{-n}(pt) = \mathbb{Z}/2$, we must appeal to a theory of modular forms over the field $\mathbb{F}_2$.

Note: $\beta(M)$ can be expressed as a polynomial in the basic form $\overline{\varepsilon} = \sum_{n \geqslant 1} q^{(2n-1)^2}$; that is $\beta(M) = a_0 + a_1 \overline{\varepsilon} + \cdots + a_m \overline{\varepsilon}^m$, where a_0 equals the Atiyah invariant $\alpha(M)$.

Here $\overline{\varepsilon}$ is the mod 2-reduction of the characteristic zero form ε. This is the only generator we need, since writing $\delta_0 = -8\delta = 1 + 24q + \cdots$, we see that $\delta_0 \equiv 1$ (mod 2). Following this line of argument [91, Theorem 3] proves that $\beta(\Omega_*^{\text{Spin}})$ is generated by $\eta, \omega\delta_0, \delta_0^2$, and ε, where η and ω were already introduced as elements in KO_1 and KO_4, respectively. ■

Part 3 of Theorem 9.1 follows as a corollary. For Part 2 [67] gives an alternative description of the subgroups $T_n(X)$ and $\widetilde{T}_n$, which also points the way to constructing a 2-local spectrum for Part 1. As a temporary measure, let G stand for the group PSp_3. If as before M is the total space of an $\mathbb{H}P^2$-bundle in the diagram:

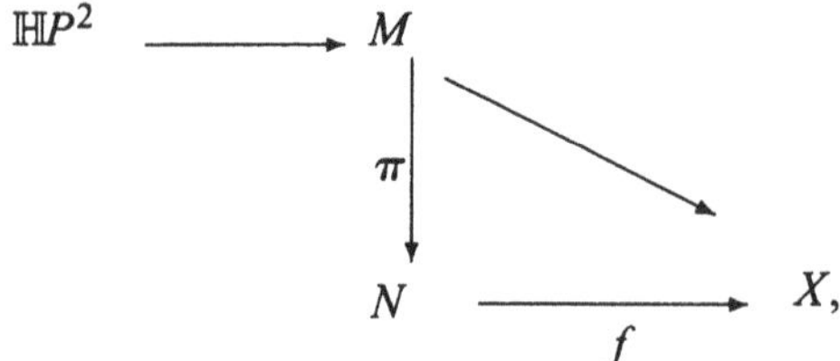

there is an obvious map:

$$\begin{gathered}\Psi : \Omega_{n-8}^{\mathrm{Spin}}(BG \times X) \to \Omega_n^{\mathrm{Spin}}(X) \\ (N, \mathrm{class}(\pi)) \times f \mapsto (M, f\pi)\end{gathered}$$

with $T_n(X)$ as the image. The subgroup $\widetilde{T}_n(X)$ is obtained by restricting Ψ to:

$$\ker(\Omega_{n-8}^{\mathrm{Spin}}(BG \times X) \xrightarrow{(pr_2)_*} \Omega_{n-8}^{\mathrm{Spin}}(X)) \cong \widetilde{\Omega}_{n-8}^{\mathrm{Spin}}(BG_\wedge X^+) \cong \widetilde{\Omega}_n^{\mathrm{Spin}}(\Sigma^8 BG_\wedge X^+).$$

Hence we have an exact sequence of (left) Ω_*^{Spin}-modules:

$$\widetilde{\Omega}_*^{\mathrm{Spin}}(\Sigma^8 BG_\wedge X^+) \xrightarrow{\widetilde{\Psi}} \Omega_*^{\mathrm{Spin}}(X) \longrightarrow \pi E_*(X) \longrightarrow 0.$$

Replace X by a point, then construct a second exact sequence by tensoring with the right exact functor $\Omega_*^{\mathrm{Spin}}(X) \otimes_{\Omega_*^{\mathrm{Spin}}} -$. This gives a commutative, exact diagram in which the right-hand vertical map is the natural transformation (♮) in Theorem 9.1, Part 2. The middle vertical map is trivially an isomorphism; the left-hand map becomes an isomorphism after inverting 2. The reason for this is that eliminating 2-torsion in the homology of $BPSp_3$ ensures that $\widetilde{\Omega}_*^{\mathrm{Spin}}(\Sigma^8 BG)$ is free over Ω_*^{Spin}; hence the left-hand map into $\widetilde{\Omega}_*^{\mathrm{Spin}}(\Sigma^8 BG_\wedge X^+)$ is a map between homology theories. Checking coefficients, i.e., taking X to be a point, shows that it — and therefore (♮) — are isomorphisms over $\mathbb{Z}[1/2]$.

9.5. Localization of πE_* at the Prime 2*

In Sec. 9.5 all abelian groups and spectra are localized at the prime 2, and as before G denotes the isometry group PSp_3 of $\mathbb{H}P^2$. In Sec. 9.4 we characterized $\pi E_*(X)$ as the cokernel of the map $\widetilde{\Psi}$, which is associated with a map of spectra:

$$M\mathrm{Spin}_\wedge \Sigma^8 BG \xrightarrow{\widetilde{T}} M\mathrm{Spin}$$

$\widetilde{T}$ emphasizes the link with the subgroup $\widetilde{T}_n$; by augmenting the domain to $M\mathrm{Spin}_\wedge \Sigma^8 G^+$, we obtain T corresponding to T_n.

Unfortunately $(\widetilde{T}_\wedge 1_X)_*$ is not injective on homotopy, so the cofiber of $\widetilde{T}$ cannot be identified with a spectrum for πE_*. We side step the problem with Proposition 9.6.

---

*This section summarizes an argument based on modern homotopy theory. We include it to give the reader some feel for elliptic homology at the prime 2.

Proposition 9.6.

1. *There exists a splitting of* $MSpin_\wedge\Sigma^8BG$ *as* $A_\vee B$ *with the property that* $\widetilde{T}|A$ *induces an injection on* $\mathbb{Z}/2$*-homology and* $\widetilde{T}|B$ *is trivial.*

2. *There is a map* $S : \bigvee\Sigma^{8k}ko \to MSpin$ *such that* $A_\vee \bigvee\Sigma^{8k}ko \xrightarrow{\widetilde{T}|A_\vee S} MSpin$ *is a homotopy equivalence.*

To define the map S for $k \geqslant 0$ let $s_k : S^{8k} \to M\text{Spin}$ be the map corresponding to the bordism class of the kth power of $\mathbb{H}P^2$ and let S be the ko-extension of the map $\vee s_k : \bigvee\limits_{k\geqslant 0} S^{8k} \to M\text{Spin}$, i.e., the composition:

$$\bigvee\Sigma^{8k}ko = ko_\wedge(\bigvee_{k\geqslant 0} S^{8k}) \to M\text{Spin}_\wedge M\text{Spin} \xrightarrow{\text{mult}} M\text{Spin}$$

If we define el as the cofiber spectrum of $\widetilde{T}|A$ and denote the projection from MSpin to el by π, then Part 2 implies Corollary 9.1.

Corollary 9.1. *The composition* $\bigvee\Sigma^{8k}ko \xrightarrow{S} MSpin \xrightarrow{\pi} el$ *is a homotopy equivalence.*

We compare the theories $\pi_*(el_\wedge X^+)$ and $\pi E_*(X)$ by means of the following commutative diagram with exact rows:

$$\begin{array}{ccccccc}
\pi_*(A_\wedge X^+) & \xrightarrow{(\widetilde{T}|A_\wedge 1)^*} & \pi_*(M\text{Spin}_\wedge X^+) & \xrightarrow{(\pi_\wedge 1)_*} & \pi_*(el_\wedge X^+) & \longrightarrow & 0 \\
\downarrow & & \| & & \downarrow & & \\
\pi_*(M\text{Spin}_\wedge\Sigma^8 BG_\wedge X^+) & \xrightarrow{(\widetilde{T}_\wedge 1)_*} & \pi_*(M\text{Spin}_\wedge X^+) & \longrightarrow & \pi E_*(X) & \longrightarrow & 0
\end{array}$$

The right-hand vertical arrow is surjective because $(\pi_\wedge 1)_*$ is surjective (Proposition 9.6). A diagram chase and the homotopy triviality of $\pi\cdot\widetilde{T}$ show that it is injective as well.

Lemma 9.1. *The composition* $\pi\cdot\widetilde{T}$ *is nullhomotopic.*

Sketch of proof: We compare π with:

$$B : M\text{Spin} \to KO[[q]],$$

the realization of the Ochanine genus β at the level of spectra. Note: The subgroup $\widetilde{T}_n(X)$ and the map β were both integrally defined, so that 2-localization has content. The spectrum $KO[[q]]$ is to be thought of as a product of countably many copies KO_j indexed by the nonnegative integers. Construction of maps $\widetilde{T}$ and B shows that their composition induces a trivial map at the level of homotopy groups. Using results from [104, 105], on MSpin-module spectra this is enough to show that $B\widetilde{T}$ itself is trivial.

Why does the same hold for $\pi\widetilde{T} = \{\pi\widetilde{T}^{(k)}\}$, with $\pi\widetilde{T}^{(k)}$ mapping into a single suspension $\Sigma^{8k}ko$ (compare Corollary 9.1)? Stolz's machine shows that it suffices to check that $\pi\widetilde{T}^{(k)}$ induces the zero map on $\mathbb{Z}/2$-homology (which is obvious from the definition of el) and that the composition with the projection $p_{8k} : \Sigma^{8k}ko \simeq ko\langle 8k\rangle \to KO$ is zero homotopic.

The triviality of $B\widetilde{T}$ implies that we have a factorization of B through el giving $\overline{B}\pi\widetilde{T}$, which is still trivial. Decompose $\overline{B}$ as $\{\overline{B}^j_k\}$ with $\overline{B}^j_k$ mapping $\Sigma^{8k}ko$ into the jth component of $KO[[q]]$.

HILFSATZ 9.1. *The map $\overline{B}^j_k : \Sigma^{8k}ko \to KO_j$ is trivial for $k > j$ and equals p_{8k} for $j = k$.*

Proof: Using the identification of el with $\bigwedge \Sigma^{8k}ko$ given by Corollary 9.1 we have

$$\Sigma^{8k}ko \xrightarrow{\;S|\Sigma^{8k}ko\;} M\text{Spin} \xrightarrow{\;B\;} KO[[q]].$$
$$\xrightarrow[\;\overline{B}^{\cdot}_k\;]{}$$

Given the definition of S in terms of the kth power of $\mathbb{H}P^2$:

$$B_* s_k = \beta(\mathbb{H}P^2)^k = \mu^k(q^k + \cdots) \in \pi_{8k}(KO)[[q]],$$

recalling that μ is the generator in degree 8 for $KO_*(pt)$. The result now follows for dimensional reasons for $\overline{B}^j_k$ and by Bott periodicity for $\overline{B}^k_k$ and the observation that $B^0 s$ agrees with p_0. ■

HILFSATZ 9.2. *The map $\pi\widetilde{T}^{(k)}$ is trivial for all values of k.*

Proof: Assume inductively that we have proved the claim for $k < j$. Then:

$$0 = \overline{B}^j_{\cdot}\pi\widetilde{T} = \sum_{0 \leqslant k} \overline{B}^j_k \pi\widetilde{T} = \sum_{0 \leqslant k \leqslant j} \overline{B}^j_k (\pi\widetilde{T})^{(k)} = p_{8k}(\pi\widetilde{T})^{(j)},$$

and so

$$(\pi\widetilde{T})^{(j)} = 0. \qquad \blacksquare$$

The fact that $\pi\widetilde{T}$ is homotopic to zero shows that $\pi E_*(X)$ agrees with $\pi_*(el_\wedge X^+)$, but it remains to construct the subspectrum A of $M\mathrm{Spin}_\wedge\Sigma^8 BG$. Once again Stolz's machine shows that it is enough to do this at the level of homology. More precisely for any ko-module spectrum Y, we let $\overline{H_*Y}$ denote the indecomposable part of H_*Y. As such it is an $A(1)_*$-comodule, where $A(1)_*$ is the Hopf algebra dual to $A(1) = \langle Sq^1, Sq^2\rangle$, since the augmentation ideal of H_*ko is an $A(1)_*$-comodule. To split

$$\overline{H_*M\mathrm{Spin}_\wedge\Sigma^8 BG}$$

it suffices to define $A = \mathrm{Image}\,(\overline{\overline{T}}_*)$ and to show that $\overline{\overline{T}}_*$ is a split surjection. We do this by imitating the steps in the proof of Proposition 9.5. We first describe the image of $\overline{S}_*$ as a $\mathbb{Z}/2$-polynomial algebra on the lowest dimensional generator of $\overline{H_*M\mathrm{Spin}}$, then show that (1) this homology coalgebra is generated by $\mathrm{Im}(\overline{S}_*)$ and $\mathrm{Im}(\overline{\overline{T}}_*)$ and (2) these images intersect trivially. This concludes our outline of the proof of Theorem 9.1, Part 1.

9.6. Introduction to the Spectrum eo_2

Let $MO\langle 8\rangle$ be the spectrum of the bordism theory of manifolds whose loop manifolds admit a spin structure, i.e., such that the classifying map of the stable tangent bundle into BO factors through the 7-connected covering space $BO\langle 8\rangle$. We already noted in Sec. 9.2 that in terms of characteristic classes, we require $w_1 = w_2 = (1/2)p_1 = 0$. Continuing from the discussion of projective plane functors, we ask whether it is possible to construct a spectrum eo_2 that bears the same relation to $MO\langle 8\rangle$ as the spectrum of connective elliptic cohomology does to $M\mathrm{Spin}$? One way of starting is with an elliptic spectrum E, whose formal group law is isomorphic to the formal completion of that of an elliptic curve defined over E^0 (point). If $1/2 \in E^0(pt)$ or if $E^*(pt)$ is torsion-free and concentrated in even dimensions, it is possible to define a structural map:

$$\sigma_E : MO\langle 8\rangle \to E,$$

which is multiplicative and modular in the sense that it commutes with elliptic-compatible maps of E. [Without the assumptions on $E^*(pt)$ take the domain spectrum to be $MU\langle 6\rangle$, a 5-connected cover of the complex bordism spectrum.]

The problem in constructing the map σ_E is that we must replace the virtual line bundle of dimension zero associated with the Hopf fiber bundle by its 3-fold tensor product to obtain a $U\langle 6\rangle$-structure. This in turn requires us to replace the symmetry condition on modular forms: $f(-1/\tau) = (-\tau)^k f(\tau)$ by a so-called cubical condition. The existence of a suitable section of a product of line bundles then guarantees the existence of σ_E [55, Theorem 6.2].

Subject to the additional technical condition of taking limits over families of elliptic curves that are (1) *étale* and (2) satisfy a higher associativity condition, the maps σ_E combine to give

$$\sigma. = MO\langle 8\rangle \longrightarrow \text{limit spectrum } eo_2.$$

This is no longer of elliptic type, but it is such that the ring $eo_2^*(pt)$ contains the ring $R_{\mathbb{Z}} = \mathbb{Z}[c_4, c_6, \Delta]/(c_4^3 - c_6^2 - 1728\Delta)$ of modular forms over $\mathbb{Z}$. Hence it is reasonable to call it the ring of *topological modular forms*.

REMARKS.

1. The associated periodic theory EO_2 has period $24^2 = 576$ obtained by inverting Δ^{24}.
2. The ring $R_{\mathbb{Z}}$ maps to the classical ring of modular forms by sending c_4 and c_6 to integral multiples of the Eisenstein series E_2 and E_3 (see Sec. 9.1).

9.7. Notes

From the definition of the corresponding genera in [53], we see that it is possible to define elliptic theories of a level greater than 2. From the point of view of [29], the need for such theories emerged as a byproduct of his study of the S^1-equivariant index of a Dirac operator on the loop space of an $O\langle 8\rangle$-manifold having coefficients in the flat bundle associated with a positive-energy representation of the central extension $\widetilde{L}\mathrm{Spin}(d)$. We discuss this type of construction in Chap. 8. (See also [29, Theorem 2.2 and Conjecture 3.9].) G. Segal has gone so far as to suggest that a geometrically based understanding of elliptic cohomology involves all levels simultaneously.

The search for geometric understanding gives significance to the 2-local theory of [67]. Elliptic objects associated with the genus zero moonshine modules are defined over $\mathbb{Z}$, and not over $\mathbb{Z}[1/2]$; indeed part of their mystery is the failure of the natural quotient character θ'_{Lg}/η_g to satisfy the genus zero condition at the prime 2. The situation may become clearer by calculating $\pi_*(el_\wedge BP^+)$ when P is a finite group of 2-power order. But in any event it is now clear that at least at level 2, elliptic (co)homology can be defined over the integers in a way that is compatible with what we already know about K-theory. The input from differential geometry is very important — construction of the spectrum el in Sec. 9.5 depends on machinery developed by Stolz to study MSpin-module spectra; this in turn was motivated by his work establishing necessary and sufficient conditions for the existence of a metric of positive scalar curvature on a smooth simply connected manifold. As with induction theorems of Brauer type used in proving Devoto's completion theorem in Chap. 7, the novice intimidated by the language of spectra has several excellent references. For a painless introduction the lectures of J. F. Adams [1, particularly Secs. 1.5–1.6 and parts of chap. 6] are recommended.

It is too early to say that the spectrum eo_2 stands in relation to Ricci curvature as el does to scalar curvature, although Stolz has raised this as a tantalizing possibility [106]. As we explain in the main text, coefficients of the theory are designed as a natural target for a genus defined on $O\langle 8\rangle$-manifolds. The discussion in Sec. 9.6 is taken from the printed version of Hopkins's lecture at the ICM (1994) in Zürich. Since it would involve yet another digression, we say nothing about cubical structures, for these it is necessary to study algebraic line bundles over abelian varieties (see references in Hopkins's paper, particularly [26]). An alternative approach involves building a spectrum eo_2 so that the integral modular forms $R_{\mathbb{Z}}$ are embedded in $\pi_*(eo_2)$. An interesting consequence of this result is a positive answer to the following prize question.

Prize question (F. Hirzebruch): Does there exist a 24-dimensional, compact, smooth manifold M with spin structure on LM such that $\hat{A}(M) = 1$ and the twisted genus $\hat{A}(M, \tau_{\mathbb{C}}) = 0$?

For such an M higher twisted $\hat{A}$-genera are closely related to dimensions of the irreducible representations of the monster $\mathbb{M}$, for example:

$$\hat{A}(M, \overline{S}^2\tau_{\mathbb{C}}) = 196883 \quad \text{and} \quad \hat{A}(M, \wedge^2\tau_{\mathbb{C}}) + \hat{A}(M, S^3\tau_{\mathbb{C}}) = 21296876.$$

Here $\overline{S}\tau_{\mathbb{C}}$ denotes the complement of a trivial line bundle in $S^2\tau_{\mathbb{C}}$.

The next prize question, at present unresolved, is to find a manifold M^{24} of the preceding kind, on which $\mathbb{M}$ acts by diffeomorphisms. Such an action would lift to the tangent bundle and also to its symmetric and exterior powers. Complexification of these bundles would have a close and natural relationship with the irreducible representations of $\mathbb{M}$, since these appear in summands of the original moonshine module V. The existence of such an action on a suitably large (in the homological sense) M^{24} is suggested by our earlier description of V as a preferred element in $Ell^*(B\mathbb{M})$.

10

$K3$-Cohomology

Chapter 10 presents some evidence for the existence of v_4-periodic cohomology theories related to complex abelian varieties of dimension 2 in the same way that the v_2-periodic theories, which we have discussed, are related to elliptic curves. We owe such theories to the work of M. Artin and B. Mazur [11, 10] on formal Brauer groups. For a topologist, however, the suggestion [67] (mentioned in Chap. 9) that replacing the pair $(\mathbb{H}P^2, PSp_3)$ by $(\mathbb{O}P^2, F_4)$ in the definition of elliptic homology may give a theory of this kind is very appealing. Perhaps by combining the two approaches, we may find a theory whose coefficients are closely related to a ring of Siegel modular forms of degree 2, defined on an open cell $H_2 \subset \mathbb{C}^3$ rather than on the upper half-plane $H_1 \subset \mathbb{C}$.

10.1. Toward a Homology Theory $\pi E_*^{\mathbb{O}}$?

Recall that we have defined

$$\pi E_*^{\mathbb{O}}(X)[\mathbb{O}P^2]^{-1} = \bigoplus_{k \geqslant 0} \Omega_{*+16k}^{\langle 8 \rangle}(X)/\sim,$$

where the equivalence relation $\sim$ is generated by identifying $[N, f] \in \Omega_*^{\langle 8 \rangle}(X)$ with $[M, f\pi] \in \Omega_{*+16}^{\langle 8 \rangle}(X)$ for an $\mathbb{O}P^2$-bundle $\pi : M \to N$, with structural group Isom $(\mathbb{O}P^2) = F_4$. To understand these groups, we must first examine the parent bordism groups more carefully.

$O\langle 8 \rangle$-bordism

The manifold M has an $O\langle 8 \rangle$-structure if $w_1 = w_2 = p_1/2 = 0$, or equivalently; if the classifying map of the stable tangent bundle $M \to BO$ factors through the

7-connected covering space $BO\langle 8\rangle$. An $O\langle 8\rangle$-structure is a homotopy class of such liftings, and bordism groups are defined in the normal way. Homotopy groups of the Thom spectrum $MO\langle 8\rangle$, give an alternative description of $\Omega_*^{\langle 8\rangle}$. We see that an $O\langle 8\rangle$-manifold is $O\langle 8\rangle$-bordant to a 7-connected manifold; hence $\Omega_n^{\langle 8\rangle}$ $(n \geqslant 16)$ equals the bordism group of n-dimensional 7-connected manifolds without specifying further tangential information.

The algebraic structure of $\Omega_*^{\langle 8\rangle}$ is partially known ([47] as well as more recent papers on the 2-torsion). We state an omnibus proposition, Proposition 10.1.

PROPOSITION 10.1. *The graded ring $\Omega_*^{\langle 8\rangle}$ has no p-torsion for $p > 3$. In dimensions less than or equal to 16, we have the following table:*

n	0	1	2	3	4	5	6	7	8
$\Omega_n^{\langle 8\rangle}$	$\mathbb{Z}$	$\mathbb{Z}/2$	$\mathbb{Z}/2$	$\mathbb{Z}/24$	0	0	$\mathbb{Z}/2$	0	$\mathbb{Z}\oplus\mathbb{Z}/2$

n	9	10	11	12	13	14	15	16
$\Omega_n^{\langle 8\rangle}$	$\mathbb{Z}/2\oplus\mathbb{Z}/2$	$\mathbb{Z}/2$	0	$\mathbb{Z}$	0	$\mathbb{Z}/2$	$\mathbb{Z}/2$	$\mathbb{Z}\oplus\mathbb{Z}$.

Torsion-free summands (detected by Pontrjagin numbers under the restriction that $p_1/2 = 0$) occur in dimensions $n \equiv 0 \pmod 4$; information about the 2-torsion can be obtained from the Adams spectral sequence for $MO\langle 8\rangle$ and about the 3-torsion by comparing $\Omega_*^{\langle 8\rangle}$ with the *framed* bordism groups [47, Sec. 4].

For one construction of the Cayley plane $\mathbb{O}P^2$, we refer the reader to Appendix B. With respect to the natural metric, the isometry group $Isom(\mathbb{O}P^2) \cong F_4$ [24], the ℓ-torsion subgroup in $H^*(BF_4,\mathbb{Z})$ is trivial for $\ell \geqslant 5$. Define the subgroups $T_n(X)$ and $\widetilde{T}_n(X)$ as before; we suppress the superscript $\mathbb{O}$ where this causes no confusion. Arguing as in [67], we have a multiplication:

$$\pi E_m(X) \times \pi E_n(Y) \to \pi E_{m+n}(X \times Y),$$

and a natural transformation:

$$\Omega_*^{\langle 8\rangle}(X) \bigotimes_{\Omega_*^{\langle 8\rangle}} \pi E_* \to \pi E_*(X), \tag{$\natural$}$$

which is compatible with multiplication on both sides.

We ask the following questions:

1. Is $\pi E_*(X) \otimes \mathbb{Z}_{(6)}$ a multiplicative homology theory?

2. Does ($\natural$) become an isomorphism after inverting 6?

3. Can we determine the coefficients $\pi E_* = \pi E_*$ (point):

a. away from 6?

b. localized at 2 and 3?

4. After localizing away from 2 and 3 and inverting the class of $\mathbb{O}P^2$, do we obtain a homology theory dual to one of the $K3$-cohomology theories defined later?

First localize away from primes 2 and 3. Taking tensor products over $\Omega_*^{\langle 8\rangle}$ and arguing as in Chap. 9, we have a commutative exact ladder:

$$\begin{array}{ccccccc}
\Omega_*^{\langle 8\rangle}(X)\otimes\widetilde{\Omega}_*^{\langle 8\rangle}(\Sigma^{16}BF_4) & \to & \Omega_*^{\langle 8\rangle}(X)\otimes\Omega_*^{\langle 8\rangle} & \to & \Omega_*^{\langle 8\rangle}(X)\otimes\pi E_* & \to 0 \\
\downarrow\wr & & \| & & \downarrow & \\
\widetilde{\Omega}_*^{\langle 8\rangle}(\Sigma^{16}BF_{4\wedge}X^+) & \longrightarrow & \Omega_*^{\langle 8\rangle}(X) & \longrightarrow & \pi E_*(X) & \longrightarrow 0
\end{array}$$

PROPOSITION 10.2. *Away from primes 2 and 3, the right-hand vertical arrow* ($\natural$) *is an isomorphism.*

Proof: Copy the earlier argument with the obvious changes. ■

To make further progress on the coefficients, we introduce maps $\alpha=\alpha^{\mathbb{O}}$ and $\beta=\beta^{\mathbb{O}}$ corresponding to the Atiyah invariant and Ochanine genus. For the first of these we have a clue on how to proceed.

Witten genus

To motivate the definition of this, we need an alternative description of the power series $Q(x)$ associated with the universal elliptic genus in Chap. 1.

PROPOSITION 10.3. *If* $g'(y)=(1-2\delta y^2+\varepsilon y^4)^{-1/2}$ *and* $Q(x)=x/f(x)$, *where* f *is the inverse power series to* g, *then:*

$$Q(x)=\frac{x/2}{\sinh(x/2)}\prod_{n=1}^{\infty}\left[\frac{(1-q^n)^2}{(1-q^ne^x)(1-q^ne^{-x})}\right]^{(-1)^n},$$

with δ *and* ε *as before.*

Proof: We follow [123, pp. 218–219].

Consider meromorphic functions $\psi : \mathbb{C} \to \mathbb{C}$ that satisfy the conditions

$$\left.\begin{array}{l} \psi(x+2\pi i\tau) = -\psi(x), \quad \psi(x+4\pi i\tau) = \psi(x), \quad \psi(-x) = -\psi(x), \\ \text{the poles of } \psi \text{ lie inside the usual lattice } L, \\ \psi(x) = \dfrac{1}{x} + O(1) \text{ as } x \to 0. \end{array}\right\} \quad (*)$$

Such a function is unique, since $\psi_1 - \psi_2$ is holomorphic and doubly periodic, hence constant and equal to zero by oddness. We sketch the proof that both $Q(x)/x$ and $1/f(x)$ satisfy $(*)$. This is not difficult for $\psi_1 = Q(x)/x$. For any such ψ, $\psi(x)^2$, and $\psi'(x)^2$ are even and invariant under L-translation, whose poles at $x = 0$ have leading terms x^{-2} and x^{-4} respectively $\pmod{L}$. Hence $\psi'(x)^2$ must be a monic quadratic polynomial in ψ^2; i.e., $(\psi')^2 = \psi^4 - 2\delta\psi^2 + \varepsilon$ for some $\delta, \varepsilon \in \mathbb{C}$. It follows that $1/\psi(x) = x + \cdots$ can be written as the inverse of a function $g(y) = y + \cdots$ given by the original elliptic integral. That δ and ε coincide with functions previously considered in Chaps. 1 and 2 follows by a suitably sophisticated comparison of coefficients. ■

Motivated by considerations in theoretical physics [120] established the following definition.

DEFINITION.

$$Q_W(x) = \frac{x/2}{\sinh(x/2)} \prod_{n=1}^{\infty} \frac{(1-q^n)^2}{(1-q^n e^x)(1-q^n e^{-x})}.$$

Label the associated genus φ_W, and for technical reasons related to modularity, evaluate it only on classes in $\Omega_*^{\langle 8 \rangle}$. This Witten genus is not quite elliptic, but it is closely related to the structural map of the cohomology theory eo_2. Proposition 10.4 is one geometric reason for regarding it as a good local candidate for the invariant $\alpha^{\mathbb{O}}$.

PROPOSITION 10.4.

1. *The simply connected closed spin manifold M^m $(m \geqslant 5)$ admits a metric of positive scalar curvature if and only if $\alpha^{\mathbb{H}}(M) = 0$.*

2. *Let M^m be the total space of an $(\mathbb{O}P^2, F_4)$-bundle over N^{m-16}. If the base manifold N admits a metric of positive Ricci curvature, then so does the total space M.*

CONJECTURE 10.1.

3. *If M^m $[m \equiv 0 \pmod 4)]$ is a smooth closed $O\langle 8 \rangle$-manifold admitting a metric of positive Ricci curvature, then $\varphi_W(M) = 0$.*

Outline proofs

N. Hitchin showed that $\alpha^{\mathbb{H}}(M)$ vanishing is a necessary condition for the existence of a metric of positive scalar curvature (see [54]). Conversely a cobordism argument shows that if $m \geqslant 5, \alpha^{\mathbb{H}}(M) = 0$ implies that the k-fold connected sum of M^m with itself admits a positive scalar curvature metric. With care we can take $m = 8$ [104]. Detailed analysis of $\pi_*(M\mathrm{Spin}_\wedge BSp_3)$ shows that for some *odd* value of k, the k-fold connected sum is spin-bordant to the total space of an $(\mathbb{H}P^2, PSp_3)$-bundle, which admits a *psc* metric because $\mathbb{H}P^2$ does. Part 1 follows by the Chinese remainder theorem. For Part 2 we assume that the bundle projection is a Riemannian submersion, from which the conclusion follows by the so-called O'Neill formulae applied to the metric on the base and the metric on the fiber, scaled by some small positive number ε. A proof of Part 3 is given in [106], provided there is an index theorem for a Dirac operator $D(LM)$ acting on sections of the spinor bundle over LM. The spinor bundle exists because LM inherits a spin structure from the $O\langle 8\rangle$-structure on M itself. We refer the reader once more to G. Segal's survey (see [101]) for further discussion.

Proposition 10.4 at least suggests how to relate $\ker \alpha^{\mathbb{O}}$ and groups $T_n^{\mathbb{O}}(pt)$ at primes other than 2 and 3. Unfortunately the existence assertion in Proposition 10.4, Part 2 is not so strong for *prc*-metrics as it is for *psc*-metrics, but the evidence so far suggests that $\alpha^{\mathbb{O}}$, i.e., the Witten genus φ_W in our assumed localization, vanishes for all (OP^2, F_4)-bundles. To show that $\ker \alpha^{\mathbb{O}} \subseteq T_n(pt)$, we must describe $O\langle 8\rangle$-generators in terms of characteristic numbers, then realize each relevant combination of characteristic numbers in terms of $\mathbb{O}P^2$-bundles. One very real stumbling block here is the absence of higher dimensional Cayley projective spaces to help us — compare [67, Sec. 4] and the proof of Proposition 9.4.

The next problem is to extend the Atiyah invariant $\alpha^{\mathbb{O}}$ to an Ochanine genus $\beta^{\mathbb{O}} : \Omega_*^{\langle 8\rangle} \to \mathbb{Q}[g_2, g_3][[q]]$, that is to the power series ring over rationalized coefficients of level 1 elliptic cohomology. We note in passing that using plumbing [53, Sec. 6.2], we can construct $O\langle 8\rangle$-manifolds mapping under the Witten genus to multiples of g_2 and g_3. Extension means that restricting $\beta^{\mathbb{O}}(M)$ to the constant term of the power series gives $\alpha^{\mathbb{O}}(M)$. In the following sections, we describe a family of cohomology theories (or more precisely a family of one-dimensional formal group laws over $\mathbb{Z}$); one of these may provide a universal $K3$-genus $\varphi^{\mathbb{O}}$, making the following diagram commute

$$
\begin{array}{ccc}
\Omega_*^{\langle 8\rangle} & \xrightarrow[\beta^{\mathbb{O}}]{} & \mathbb{Q}[g_2, g_3][[q]] \\
{\scriptstyle \varphi^{\mathbb{O}}}\big\downarrow & & \big\downarrow \\
\text{Modular forms defined on } H_2 \subset \mathbb{C}^3 & \xrightarrow[i]{} & \mathbb{Q}[[q', q]].
\end{array}
$$

Here the right-hand vertical arrow expands g_2 and g_3 in terms of the second variable q', and the inclusion map i first restricts a Siegel modular form to a subspace isomorphic to $H_1' \times H_1$ in H_2, then takes the power series expansions. If such a map $\beta^{\mathbb{O}}$ can be constructed, the final problem is to show that $\ker \beta^{\mathbb{O}} = \widetilde{T}_n$ in dimension n and to determine its image. The image depends on values taken by $\alpha^{\mathbb{O}}$on low-dimensional manifolds and the arithmetic of modular forms. As in [67] it remains to decompose $\Omega_n^{\langle 8\rangle}$ as the sum of $\widetilde{T}_n$ and a submodule coming from combinations of a finite family of manifolds M^m in low dimensions previously used to describe $\operatorname{Im}\beta^{\mathbb{O}}$.

Localization at 2 and 3

As in the case of elliptic cohomology and $\mathbb{H}P^2$-bundles, it follows from the definition of $\widetilde{T}_n$ that there is an exact sequence of (left) $\Omega_*^{\langle 8\rangle}$-modules:

$$\widetilde{\Omega}_*^{\langle 8\rangle}(\Sigma^{16}BF_4{}_\wedge X^+) \underset{\widetilde{Z}}{\longrightarrow} \Omega_*^{\langle 8\rangle}(X) \longrightarrow \pi E_*^{\mathbb{O}}(X) \longrightarrow 0.$$

Passing to the homotopy groups of spectra, $\widetilde{Z}$ can be written as $(\widetilde{T}_\wedge 1)_*$: $\pi_*(MO\langle 8\rangle_\wedge \Sigma^{16}BF_4{}_\wedge X^+) \to \pi_*(MO\langle 8\rangle_\wedge X^+)$, which suggests, as in Sec. 9.5, that the cofiber of $\widetilde{T}$ is closely related to a spectrum corresponding to $\pi E_*^{\mathbb{O}}$. More precise information depends on 2- and 3-local definitions of $\alpha^{\mathbb{O}}$ and $\beta^{\mathbb{O}}$; these in turn depend on knowing $\Omega_{n,p}^{\langle 8\rangle}$ for $p = 2,3$. Some information in given in [47], and in principle the answer is known for the prime 3. For the prime 2 less is known, since progress depends on understanding differentials in the Adams spectral sequence for the 2-primary part of $\pi_*(MO\langle 8\rangle)$. Pushing these calculations further is a preliminary to answering Questions 1 and 3b.

The situation for Questions 3a and 4 is more promising since there are candidates for the dual cohomology theory. We now turn to the construction of these.

10.2. Abelian Varieties

This section contains a utilitarian overview of part of a vast subject — our references are [100, 74], and [32] for the special case of Jacobians of curves of genus 2.

Let V denote a complex vector space of dimension g and let L be a lattice in V acting as usual on V by addition. The quotient space V/L is called a *complex torus*, and it has the structure of a connected, compact, complex Lie group.

A *Riemann form* on X is an Hermitian form H on V such that the imaginary part $E = \operatorname{im} H$ is integer-valued on L, i.e., $E(\ell_1, \ell_2) \in \mathbb{Z}$ for all $\ell_1, \ell_2 \in L$. If

$H(v,v) \geqslant 0$ for all $v \in V$, H is a *positive* Riemann form. If in addition $H(v,v) > 0$ for all $v \neq 0$, H is nondegenerate.

DEFINITION. The complex torus $X = V/L$ is an abelian variety over $\mathbb{C}$ of dimension g if X possesses a positive, nondegenerate Riemann form.

It is not difficult to see that if $(H, E = \operatorname{im} H)$ is a Riemann form, then $E(iv, iw) = E(v, w)$ for all $v, w \in V$. Furthermore the existence of an alternating form E satisfying this condition, and the earlier one that $E(L, L) \subseteq \mathbb{Z}$, is equivalent to the existence of a holomorphic line bundle ξ over X with Chern class $c_1(\xi)$ represented by E (see [74, Chap. 2, Sec. 1]). Conversely given the bundle ξ, we can use c_1 to define E and the restrictions on E to construct the Hermitian form H. Using the bundle associated with the Riemann form, it is possible to prove that X admits an embedding in some high-dimensional $\mathbb{C}P^n$ (Lefschetz Theorem, [74, Chap. 4, Theorem 5.1]), justifying the use of the label abelian variety for the pair (X, H).

REMARK. For $g \geqslant 2$ there are complex tori that are not abelian varieties. The space of $\mathbb{C}$-tori of dimension g depends on g^2 complex parameters, but as we sketch later, that of g-dimensional abelian varieties depends on $g(g+1)/2$ parameters (see [100, p. 99] for a specific example).

A *polarized* abelian variety is a torus X and an equivalence class of positive Riemann forms containing a nondegenerate representative H. The pairs (V, H_1) and (V, H_2) are equivalent if there exist natural numbers n_1, n_2 with $n_1 H_1 = n_2 H_2$. Polarization is *principal* if for some H in the class, we can choose a (symplectic) basis for the lattice L such that E is given by the matrix:

$$\begin{pmatrix} 0 & 1_g \\ -1_g & 0 \end{pmatrix}.$$

Almost by definition every polarized abelian variety is isogenous to a principally polarized abelian variety (see [100, p. 96]).

Examples of principal polarizations are given by Jacobian varieties of nonsingular algebraic curves. One very classical way of defining Jac (C) for a curve of genus g is to map the first homology group $H_1(C, \mathbb{Z})$ into $\mathbb{C}^g$, then define Jac (C) to be the complex torus $\mathbb{C}^g / H_1(C, \mathbb{Z})$.

At least locally near p_0 map p to $\mathbb{C}^g$ by $p \mapsto \left(\int_{p_0}^{p} \omega_1, \ldots, \int_{p_0}^{p} \omega_g\right)$, where the holomorphic differential $\omega_i = z^{i-1} dz / \sqrt{f(z)}$ for $i = 1, 2, \ldots g$. (The function $f(z) = (z - a_1) \cdots (z - a_d)$ with $g = [(d-1)/2]$ and constants a_j pairwise unequal). To extend this definition to all of C, we must work modulo the values of integrals along all possible closed paths, i.e., take values in $\mathbb{C}^g / H_1(C, \mathbb{Z})$, where the image of a basic cycle γ in H_1 equals $(\int_\gamma \omega_1, \ldots, \int_\gamma \omega_g)$.

In the special case when $g = 2$ (which is actually the only one of interest), the image $\alpha(C)$ in Jac (C) carries a two-dimensional homology class whose de Rham dual is represented by an alternating form E over the real numbers $\mathbb{R}$. In turn the form E can be interpreted as the imaginary part of a polarization H; this turns out to be principal. This is not entirely unexpected, since we start with a complex structure on a 2-manifold for which there is a natural basis for one-dimensional homology having intersection matrix $\left(\begin{smallmatrix} 0 & 1_2 \\ -1_2 & 0 \end{smallmatrix}\right)$. For curves of higher genus, the map $\alpha : C \to \text{Jac }(C)$ induces a map $\alpha^{(g-1)}$ of the symmetric product of C with itself $(g-1)$ times into the Jacobian, whose cohomological dual again determines the canonical principal polarization. In general a moduli space for a set of abelian varieties with some additional structure means a complex analytic space or manifold, whose points are naturally bijective with the varieties concerned.

We are particularly interested in the moduli space of principally polarized abelian varieties, which generalizes the quotient of the upper half-plane $H_1 \subset \mathbb{C}$ by $SL_2(\mathbb{Z})$ already discussed in Chap. 2. Let $X = V/L$ be such a variety, and choose a $\mathbb{C}$-basis for V and a symplectic basis for the lattice L. In terms of these coordinates $X \cong \mathbb{C}^g/\langle\omega_1, \ldots, \omega_{2g}\rangle$, where $\omega_1, \ldots, \omega_{2g}$ are $\mathbb{R}$-linearly independent column vectors and the imaginary part of the accompanying Riemann form has matrix $E(\omega_i, \omega_j) = \left(\begin{smallmatrix} 0 & 1_g \\ -1_g & 0 \end{smallmatrix}\right) = J$. The $g \times 2g$ complex matrix $\Omega = (\omega_1, \ldots, \omega_{2g})$ is called the period matrix.

PROPOSITION 10.5. *The period matrix Ω determines an abelian variety if and only if the following two conditions (the Riemann relations) hold*

***R*1:** $\Omega J \Omega^T = 0$

***R*2:** $2i(\overline{\Omega} J^{-1} \Omega^T)^{-1}$ *is positive definite.*

Proof: The relation $R1$ is equivalent to the condition $E(iv, iw) = E(v, w)$.

With respect to the standard basis of $\mathbb{C}^g$, the Riemann Hermitian form $E(iv, w) + iE(v, w)$ equals $2i(\overline{\Omega} J^{-1} \Omega^T)^{-1}$, which gives $R2$.

For more details, see [74, pp. 75–77]. ■

DEFINITION. Let H_g be the space of $g \times g$ complex matrices τ that are symmetric and have positive definite imaginary part; H_g is a complex manifold of dimension $g(g+1)/2$ (the Siegel upper half-space), which admits an action by $Sp_{2g}(\mathbb{R})$.

$$\text{If } \begin{pmatrix} A & B \\ C & D \end{pmatrix} \in Sp_{2g}(\mathbb{R}), \text{ then } \tau \mapsto (A\tau + B)(C\tau + D)^{-1}.$$

We recall that the symplectic condition implies that A^TC and B^TD are symmetric and $A^TD - C^TB = 1_g$.

PROPOSITION 10.6. *The moduli space of isomorphism classes of principally polarized abelian varieties of dimension g is isomorphic to* $Sp_{2g}(\mathbb{Z})\backslash H_g$.

Proof: Write the period matrix Ω as (Ω_1, Ω_2), so that the Riemann relations take the form:

R1′: $\Omega_2\Omega_1^T - \Omega_1\Omega_2^T = 0$.

R2′: $2i(\Omega_2\overline{\Omega}_1^T - \Omega_1\overline{\Omega}_2^T) > 0$.

Both Ω_1 and Ω_2 are invertible, and the set of all pairs satisfying $R1'$ and $R2'$ admits a left action by $GL_g(\mathbb{C})$ and a right action by $Sp_{2g}(\mathbb{Z})$. The first action gives an isomorphism of abelian varieties, the latter a change of symplectic basis for the lattice L. Hence as far as moduli go, we can replace (Ω_1, Ω_2) by $(\tau, 1_g) = (\Omega_2^{-1}\Omega_1, 1_g)$. Condition $R1'$ implies that τ is symmetric, condition $R2'$ that its imaginary part is positive definite. If $(\tau, 1_g) \sim (\tau', 1_g)$ under the symplectic action, then:

$$\begin{aligned}(\tau, 1_g)\begin{pmatrix} A & B \\ C & D \end{pmatrix} &= (\tau A + C, \tau B + D) \\ &\sim ((\tau B + D)^{-1}(\tau A + C), 1_g) \\ &= (\tau', 1_g).\end{aligned}$$

In the preceding definition we noted that $\tau = \tau^T$ and rewrote the action in a more familiar form. ■

When $g = 2$ we obtain Proposition 10.7.

PROPOSITION 10.7. *A principally polarized abelian surface is either the Jacobian of a smooth curve of genus 2 or the canonically polarized product of two elliptic curves.*

Proof: By Teichmüller theory the moduli space of complex structures on a curve of genus 2 has dimension 3, equal to that of $Sp_4(\mathbb{Z})\backslash H_2$. By Torelli's theorem [74, Chap. 11, Theorem 1.7 and Exercise 8], if two canonically polarized Jacobians are isomorphic, so are the original curves. Hence, allowing for products $C_1 \times C_2$, the Jacobian construction exhausts the moduli space of surfaces. ■

REMARKS. A similar result holds for abelian 3-folds. We note that $X = \mathbb{C}^2/L$ splits as a product of elliptic curves if L splits as $L_1 \oplus L_2$ with each factor isotropic with respect to the form $U^T \left(\begin{smallmatrix} 0 & 1 \\ -1 & 0 \end{smallmatrix}\right) V$ (see [74, p. 313]). Interestingly enough any abelian surface whose endomorphism structure is of definite quaternionic type over $\mathbb{Q}$ is the product of an elliptic curve with complex multiplication with itself (see [74, pp. 267 and 314]).

Proposition 10.7 suggests that if there is a cohomology theory where $\mathbb{O}P^2$ plays the same role as $\mathbb{H}P^2$ in elliptic cohomology, then its formal group law must be encoded in the geometry of Jac (C), with C a generic curve of genus 2. For this geometry we refer the reader to [32].

DEFINITION. Let $X = V/L$ be an abelian variety of dimension g. The *Kummer variety* associated with X is defined to be the quotient space $X/\langle(-1)_X\rangle$ obtained by identifying x with $-x$ for all $x \in X$.

The quotient K_X is an algebraic variety of dimension g over $\mathbb{C}$, which is smooth away from 2^{2g} singular points of multiplicity 2^{g-1}, the images under the projection map of the 2-division points of X. Assuming that X comes with (irreducible) principal polarization, there is an embedding:

$$\psi : K_X \to \mathbb{C}P^{2^g-1}.$$

Taking $g = 2$ we therefore obtain an embedding of an algebraic surface with 16 singular points in $\mathbb{C}P^3$. The picture can be described quite explicitly as follows:

Let

$$y^2 = F(x) = f_0 + f_1 x + \cdots + f_6 x^6 \in \mathbb{C}[x]$$

describe a curve of genus 2. We assume that f has no multiple factors. It can be shown [32, Chap. 1] that every curve of genus 2 defined over $\mathbb{C}$ is birationally equivalent to a curve of this kind, so that our model is generic. By first constructing the Jacobian, then dividing out by the involution, it is possible to obtain the locus of the Kummer surface in $\mathbb{C}P^3$ in the form:

$$K(\xi_1, \ldots, \xi_4) = K_2 \xi_4^2 + K_1 \xi_4 + K_0 = 0, \tag{10.1}$$

where:

$$\begin{aligned}
K_2 &= \xi_2^2 - 4\xi_1\xi_3, \\
K_1 &= -2\{2f_0\xi_1^3 + f_1\xi_1^2\xi_2 + 2f_2\xi_1^2\xi_3 + f_3\xi_1\xi_2\xi_3 + 2f_4\xi_1\xi_3^2 + f_5\xi_2\xi_3^2 + 2f_6\xi_3^3\}, \\
K_0 &= (f_1^2 - 4f_0f_2)\xi_1^4 - 4f_0f_3\xi_1^3\xi_2 - 2f_1f_3\xi_1^3\xi_3 - 4f_0f_4\xi_1^2\xi_2^2 \\
&\quad + 4(f_0f_5 - f_1f_4)\xi_1^2\xi_2\xi_3 + (f_3^2 + 2f_1f_5 - 4f_2f_4 - 4f_0f_6)\xi_1^2\xi_3^2 \\
&\quad - 4(f_0f_5\xi_1\xi_2^3) + 4(2f_0f_6 - f_1f_5)\xi_1\xi_2^2\xi_3 + 4(f_1f_6 - f_2f_5)\xi_1\xi_2\xi_3^2 - 2f_3f_5f_1\xi_3^3 \\
&\quad - 4f_0f_6\xi_2^4 - 4f_1f_6\xi_2^3\xi_3 - 4f_2f_0\xi_2^2\xi_3^2 - 4f_3f_6\xi_2\xi_3^3 + (f_5^2 - 4f_4f_6)\xi_3^4.
\end{aligned}$$

The geometry of this surface is beautiful. The vertex $\mathbf{0} = (0,0,0,1)$ of the simplex of reference in $\mathbb{C}P^3$ is a *node*; i.e., in the neighborhood of $\mathbf{0}$ the surface behaves like the cone $K_2 = 0$. The same holds for the 15 other images of the 2-division points on X. Dual to the nodes are the *tropes*, 16 singular tangent planes with the property of touching the surface along a conic instead of at a single point. Every node is incident with precisely six tropes and vice versa. There are six tropes through $\mathbf{0}$ given by equations

$$\theta_i^2 \xi_1 - \theta_i \xi_2 + \xi_3 = 0 \qquad (1 \leqslant i \leqslant 6),$$

where θ_i are zeros of the polynomial $F(x)$, and these are sufficient to solve the inverse problem of describing a curve C such that Jac $(C)/\{\pm 1\}$ coincides with a *given* Kummer surface K.

We assume K has 16 nodes and 16 tropes, with respect to a suitable system of coordinates, $\mathbf{0} = (0,0,0,1)$ is a node, and the tangent cone of this node is nondegenerate. Under these conditions it follows that K is defined by an equation similar to Eq. (10.1), that is $K(\xi_1, \xi_2, \xi_3, \xi_4) = K_2 \xi_4^2 + K_1 \xi_4 + K_0$ with $K_2 = \xi_2^2 - 4\xi_1 \xi_3$.

Proposition 10.8. *Under the assumptions just made there is a constant c such that K corresponds to the algebraic curve $y^2 = F(x)$ with*

$$F(x) = \frac{1}{c}(K_1^2 - 4K_2K_0).$$

The right-hand side is evaluated at $(\xi_1, \xi_2, \xi_3) = (0, 1, x)$.

Proof: In the generic situation there are 6 tropes through $\mathbf{0}$, tangent to the cone $K_2 = 0$ at the points $(1, 2\theta_i, \theta_i^2)$ for $1 \leqslant i \leqslant 6$. The trope T_i has dual coordinates $(\theta_i^2, -\theta_i, 1)$. Any line in a trope meets the surface K only in double points. Hence the discriminant of the defining equation with respect to ξ_4 vanishes on all six tropes. Comparing degrees we must have

$$K_1^2 - 4K_2K_0 = c = \prod_{i=1}^{6} (\theta_i^2 \xi_1 - \theta_i \xi_2 + \xi_3).$$

Now set $\xi_1 = 0$, $\xi_2 = 1$, $\xi_3 = x$, obtaining the required rational expression:

$$\prod_{i=1}^{6} (x - \theta_i) = \frac{1}{c}(K_1^2 - 4K_2K_0) = F(x). \qquad \blacksquare$$

It can be shown that the configuration of 16 points and 16 planes with the incidence properties of the nodes and tropes of a Kummer surface is determined (up to an elementary collineation) by six planes tangent to a quadric cone. Hence the given surface is projectively equivalent to that constructed as in [32], starting from $y^2 = F(x)$.

Kummer surfaces embedded in $\mathbb{C}P^3$ that we consider are examples of a more general class of $K3$-surfaces Y. In the language of algebraic geometry, these are characterized by the following properties, which imply that a $K3$-surface normally embedded in $\mathbb{C}P^n$ has degree $2n-2$. We are interested in the special case $n=3$:

***K*1:** The irregularity $h^{1,0}(Y)$ (equal to the number of holomorphic 1-forms) equals 0.

***K*2:** The canonical bundle K_Y (equal to the second exterior power of the holomorphic cotangent bundle) is trivial. This condition implies that Y admits a nowhere vanishing holomorphic 2-form ω.

Let Y be a quartic surface of the type we are considering. Since $H^1(\mathbb{C}P^3, \mathbb{Q}) = 0$, the same holds for Y; therefore it satisfies $K1$. Since Y has degree 4 and $K_{\mathbb{C}P^3}$ equals the class of $(-4)\times$ canonical line bundle, the adjunction formulae (see for example [50, pp. 146–147]) imply that $K_Y \equiv 0$.

For a general $K3$-surface the Hirzebruch–Riemann–Roch formula implies that evaluating the second Chern class on the fundamental class gives 24 — this determines the second Betti number $\beta_2 = 22$. Using duality we have the so-called Hodge diamond of dimensions $h^{i,j}$ of the cohomology groups $H^{i,j}$:

$$\begin{array}{ccccc} & & 1 & & \\ & 0 & & 0 & \\ 1 & & 20 & & 1. \\ & 0 & & 0 & \\ & & 1 & & \end{array}$$

Note $h^{1,0} = h^{3,0} = 0$ and $h^{2,0} = 1$.

10.3. *K*3-Cohomology

We encountered formal groups in earlier chapters. Abstractly an n-dimensional formal group over the ring K is a functor G from the category of nilpotent K-algebras to abelian groups, whose underlying set-valued functor admits an n-dimensional affine coordinatization. The simplest examples are the one-dimensional additive and multiplicative formal groups $\mathbb{G}_a^\wedge$ and $\mathbb{G}_m^\wedge$. We choose coordinates for the latter so that:

$$G(x_1, x_2) = x_1 + x_2 - x_1x_2,$$

which is equivalent to taking its logarithm:

$$g(y) = \sum_{n=1}^{\infty} \frac{y^n}{n} = -\log(1-y).$$

Both laws are defined over $\mathbb{Z}$.

Given a sheaf $\mathfrak{A}$ of abelian groups over a variety X, for example the structural sheaf $\mathcal{O}_X$, and a formal group law, we can apply the sheaf construction to the latter starting with the presheaf $\Gamma(U, G(\mathfrak{A})) = G(\Gamma(U, \mathfrak{A}))$. In this way we obtain $\mathbb{G}^{\wedge}_{m,X} := \mathbb{G}^{\wedge}_{m,\mathcal{O}_X}$.

EXAMPLE. If X is an algebraic curve, the Picard group $\mathrm{Pic}(X)$ equals the identity component of $H^1(X, \mathbb{G}_{m,X})$.

Can we say something similar about higher cohomology groups of more general varieties? More precisely under what conditions do groups $H^i(X, \mathbb{G}^{\wedge}_{m,X})$ have a formal group structure if X is defined over an algebraically closed field? The general answer is contained in [11] (see also [10] for the special case $i = 2$). Under conditions $H^{i-1}(X, \mathcal{O}_X) = H^{i+1}(X, \mathcal{O}_X) = 0$, we obtain a formal group of dimension $h^{i,0} = \dim H^i(X, \mathcal{O}_X)$. When $i = 2$, we write $H^2(X, \mathbb{G}_{m,X}) := \widehat{Br}(X)$; the formal Brauer group of the variety X; for a $K3$-surface, it is one-dimensional. Artin shows that in characteristic p, either $\widehat{Br}(X)$ is p-divisible of height $h (1 \leqslant h \leqslant 10)$ or $h = \infty$ and $\widehat{Br}(X) \cong \mathbb{G}^{\wedge}_{a,X}$. Leaving aside this last (supersingular) case, we can show that the height induces a stratification of the $K3$-moduli space and this is *completely regular*. We can now apply Landweber's exactness criterion, then define $K3$-cohomology by pulling back the sheaf of modules over the space of one-dimensional formal groups (each the homomorphic image of the universal example in cobordism) to the $K3$-moduli space. For a cross reference see [102, Sec. 4], particularly Proposition 14 on the inductive construction of Cohen–Macaulay modules.

REMARK. Strictly speaking we require Landweber's criterion reduced modulo p. Alternatively we can work over an arithmetic moduli space of $K3$-surfaces.

Rather than go through this general argument, we restrict attention to the special case of X equal to a hypersurface of degree d embedded in $\mathbb{C}P^{d-1}$, considered by [103]. For this we are able to obtain a formal group with coordinates defined over $\mathbb{Z}$ and hence an explicit logarithm.

THEOREM 10.1. *Let K be a Noetherian ring and F a homogeneous polynomial of degree $d = N+1$ in $K[t_0, \ldots, t_N]$. Let X be the subscheme of the projective space KP^N defined by the ideal $\langle F \rangle$, assumed to be flat over K. Then $H^{N-1}(X, \mathbb{G}^{\wedge}_{m,X})$ is a one-dimensional formal group over K. If in addition K is flat over $\mathbb{Z}$, then the* *logarithm of the formal group law is given by:*

$$g(y) = \sum_{m \geqslant 1} m^{-1} \beta_m y^m$$

with β_m *equal to the coefficient of* $(t_0 t_1 \cdots t_N)^{m-1}$ *in* $(F)^{m-1}$.

REMARKS. We state the result for a Noetherian ring K rather than for the complex numbers $\mathbb{C}$, to emphasize that the formal group law associated with an integrally defined Kummer surface has a logarithm in $\mathbb{Q}[[y]]$. For the sake of clarity and also because our main interest is the quartic Kummer surface, we simplified Stienstra's assumptions. He actually proves the analog of Theorem 10.1 for subschemes defined by a family of polynomials $F_1,\ldots,F_r$, with d equal to the sum of degrees and $n = \binom{d-1}{N}$, giving an n-dimensional formal group.

Proof: The proof of Theorem 10.1 is as follows:

STEP 1: Let K be a ring $f : X \to Y$, an affine morphism of schemes over K with X Noetherian and flat over K. Then for every $i \geqslant 0$, there is a functorial isomorphism $H^i(X, \mathbb{G}^\wedge_{m,X}) \cong H^i(Y, \mathbb{G}^\wedge_{m,f_*\mathcal{O}_X})$.

The proof uses standard arguments from sheaf-theoretic cohomology [103, Secs. 3.2–7]. Here and elsewhere we use *small* extensions to carry out inductive arguments. A small extension is a surjection of nilpotent K-algebras with the kernel generated as a K-module by a single nonzero element ε with $\varepsilon^2 = 0$.

STEP 2: Abusing notation let $\langle F \rangle$ denote the ideal sheaf on KP^N associated with the ideal of $K[t_0,\ldots,t_N]$ generated by the homogeneous polynomial F of degree $N+1$. By Step 1 the inclusion $f : X \to KP^N$ induces an isomorphism:

$$H^{N-1}(X, \mathbb{G}^\wedge_{m,X}) \cong H^{N-1}(KP^N, \mathbb{G}^\wedge_{m,f_*\mathcal{O}_X}).$$

A dimension-shifting argument (more straightforward than that in [103, Sec. 4] because of our assumptions) shows that the right-hand side is isomorphic to $H^N(KP^N, \mathbb{G}^\wedge_{m,\langle F \rangle})$. We must find a formal group law for this group with coordinates.

STEP 3: $H^N(KP^N, \mathbb{G}^\wedge_{m,\langle F \rangle})$ is isomorphic to the Čech group $\check{H}^N(\mathfrak{U}, \mathbb{G}^\wedge_{m,\langle F \rangle})$, where $\mathfrak{U}$ is the standard open cover of $KP^N, \mathfrak{U} = \{U_0,\ldots,U_N\}$, with $t_i \neq 0$ on U_i.

STEP 4: Use the coordinatization $G(x_1,x_2) = x_1 + x_2 - x_1 x_2$ for $\mathbb{G}^\wedge_m$, for every nilpotent K-algebra A. We identify the sheaf $\mathbb{G}^\wedge_{m,\langle F \rangle}(A)$ with $\langle F \rangle \underset{K}{\otimes} A$, inheriting its formal group law from G. Choose coordinates for $\check{H}^N(\mathfrak{U}, \mathbb{G}^\wedge_{m,\langle F \rangle})$ by mapping

$a \in A$ to the Čech N-cocycle given by $F(t_0 t_1 \cdots t_N)^{-1} \otimes a$ on $U_0 \cap \cdots \cap U_N$. This map is a functorial isomorphism by induction over algebras built by a sequence of small extensions. The induction starts, since the sheaf $\langle F \rangle$ is isomorphic to $\mathcal{O}_{KP^N}(-(N+1))$, and for algebras of the type $\{K\varepsilon : \varepsilon^2 = 0\}$ the bijection:

$$K\varepsilon \longrightarrow \check{H}^N(\mathfrak{U}, \mathbb{G}^{\wedge}_{m,\langle F \rangle}(K\varepsilon)) = \check{H}^N(\mathfrak{U}, \langle F \rangle \underset{K}{\otimes} K\varepsilon)$$

is a consequence of the known cohomology of $\langle F \rangle$, identified above. This initial isomorphism establishes the dimension of the formal group law as 1 [more generally $n = \binom{d-1}{N}$]. For a general algebra A, we take a small extension:

$$0 \to K\varepsilon \to A \to \overline{A} \to 0,$$

and choose coordinates for $\check{H}^N(\mathfrak{U}, \mathbb{G}^{\wedge}_{m,\langle F \rangle}(A))$ using the result above for $K\varepsilon$ and an assumed result for the quotient $\overline{A}$. This involves a diagram chase [103, Sec. 4.9].

STEP 5: The formal group law on the tensor product $\langle F \rangle \underset{K}{\otimes} A$ explains the equation for the logarithm. Powers of the term $F(t_0 t_1 \cdots t_N)^{-1} \otimes y$ combine to give $\sum_{m \geqslant 1} F^m(t_0 t_1 \cdots t_N)^{-m} \otimes (y^m/m)$, hence:

$$g(y) = \sum_{m \geqslant 1} m^{-1} \beta_m y^m$$

with β_m as stated in Theorem 10.1. Note: We made three choices of coordinates — in the defining equation for X, the Čech N-cocycle $F(T_0 \cdots t_N)^{-1}$, and the coordinatization of the multiplicative group law $\mathbb{G}^{\wedge}_m$. The algorithm shows how these choices affect the final result for $H^{N-1}(X, \mathbb{G}^{\wedge}_{m,X})$. ■

EXAMPLE. Let X be the Fermat hypersurface in $\mathbb{Z}P^{d-1}$ defined by the equation $t_0^d + \cdots + t_{d-1}^d = 0$. The formal group structure on $H^{d-2}(X, \mathbb{G}^{\wedge}_{m,X})$ has logarithm:

$$g(y) = \sum_{m \geqslant 0} \frac{(dm)!}{(m!)^d} \frac{y^{dm+1}}{dm+1}.$$

This follows from inspecting the iterated binomial expansion of $(t_0^d + \cdots + t_{d-1}^d)^{dm}$.

10.4. Siegel Modular Forms and Open Questions

Let us again consider the moduli space $Sp_4(\mathbb{Z}) \backslash H_2$ of equivalence classes of principally polarized abelian surfaces. Topologically H_2, the set of 2×2 complex symmetric matrices with a positive definite imaginary part, is an open convex subset of $\mathbb{C}^3$. We introduce holomorphic functions as complex-valued functions that can be represented locally by power series.

DEFINITION. A Siegel modular form of degree 2 and weight k is a holomorphic function $f : H_2 \to \mathbb{C}$ such that for each matrix $\left(\begin{smallmatrix} A & B \\ C & D \end{smallmatrix}\right) \in Sp_4(\mathbb{Z})$ and all $\tau \in H_2$ $\;f(\tau) = \det(C\tau + D)^{-k} f((A\tau + B)(C\tau + D)^{-1})$.

For the sake of completeness, we also require f to be bounded on any subset $\{\tau \in H_2 : \operatorname{Im}(\tau) \geqslant c \text{ with } c > 0\}$. Although this condition is redundant by a theorem of M. Koecher for the full modular group, the analog fails for H_1, and there may be problems when $Sp_4(\mathbb{Z})$ is replaced by a congruence-related subgroup [65, Theorem 4.1].

Examples of modular forms of degree 2 are provided by the Eisenstein series. If $\Gamma_2 = Sp_4(\mathbb{Z})$, let $\Gamma_{2,0}$ denote the subgroup whose the submatrix $C = 0$. Let M denote a member of a family of coset representatives for the quotient space $\Gamma_{2,0} \backslash \Gamma_2$. Then for all even positive integers k, we define

$$E_k(\tau) := \sum_{M \in \Gamma_{2,0} \backslash \Gamma_2} \det(C\tau + D)^{-k}.$$

For a discussion of the convergence of Eisenstein series in degree 2, see [65, Theorem 5.1]. We next form the graded ring $M_* = \bigoplus_{k \in \mathbb{Z}} M_k$, where M_k denotes the vector space over $\mathbb{C}$ of forms of weight k. If $f_k \in M_k$ and $f_l \in M_l$, then the product is given by:

$$\begin{gathered} M_k \times M_l \to M_{k+l} \\ (f_k, f_l) \mapsto f_k f_l \end{gathered}$$

As a $\mathbb{C}$-algebra M_* is finitely generated; we already used the corresponding result for forms of degree 1, taking the elliptic invariants g_2 and g_3 (which are essentially Eisenstein forms of weights 4 and 6) as generators. In degree 2 we explain how M_{even} is generated by the functions E_4, E_6, E_{10}, and E_{12}. This result can be extended to include odd weights by adding an extra generator with $k = 35$. Note: The boundedness condition implies that $M_k = 0$ for negative values of k and a form of weight zero is constant. We start with a theta-type function.

DEFINITION. Let $\tau \in H_2$ and a, b be column vectors from $\mathbb{Z}^2$. A theta series of characteristic (a, b) is given by the equation:

$$\vartheta(\tau; a, b) = \sum_{z \in \mathbb{Z}^2} e^{\pi i(\tau[z + (a/2)] + b^T z)},$$

where $\tau[z + (a/2)] = (z + (a/2))^T \tau (z + (a/2))$.

Table 10.1. Theta Functions

a	00	01	10	11	00	01	00	10	00	11
b	00	00	00	00	10	10	01	01	11	11

Reducing the characteristic (a,b) modulo 2 gives the possibilities displayed in Table 10.1. Write $\Theta(\tau) = \prod_{\substack{(a,b)\\ \text{mod}\, 2}} \vartheta(\tau;a,b)$.

Direct calculation [65, Sec. 9] gives

1. $\Theta^2(\tau)$ is a modular form of weight 10. $\Theta(\tau)$ itself is almost modular of weight 5, with a twisting given by a nontrivial character of Γ_2.
2. Modulo the action of Γ_2, the zeros of Θ, all of which have order 1, lie in the subset $N = \{\tau \in H_2 : \tau = \text{diag}(z_1, z_4)\}$ isomorphic to $H_1 \times H_1$.

A corollary of the calculation is that any modular form of even weight that vanishes on N is divisible by Θ^2 in the ring of holomorphic functions on H_2.

In degree 1, up to scalar multiples, elliptic invariants g_2 and g_3 coincide with the classical Eisenstein series e_4 and e_6. As a temporary measure we use the lower case notation to distinguish degree 1 from degree 2.

PROPOSITION 10.9. *Let $F \in M_k$ (k = even) be a modular form of degree 2. Then the restriction of F to the subset N of H_2 can be represented as an isobaric polynomial in the three functions:*

$$e_4(z_1)e_4(z_4), \qquad e_6(z_1)e_6(z_4), \qquad e_4^3(z_1)e_6^2(z_4) + e_4^3(z_4)e_6^2(z_1).$$

Here isobaric means with respect to the weights 4, 6, and 12.

Proof: Observe that $F|_N(z_1, z_4)$ is a degree 1 modular form in each variable z_1, z_4 if the other is held fixed. For degree 1 forms we already know that the functions e_4 and e_6 suffice, and the result follows by calculation. ∎

Restricting E_4 and E_6 to N immediately gives the first two functions. The restriction of E_{12} must be a linear combination of E_4^3, E_6^2, and the third function in Proposition 10.9. If $E_{12} - \alpha E_4^3 - \beta E_6^2$ vanishes on N, repeated division by Θ^2 shows that it must be a modular form of negative weight on H_2, hence zero. Comparing Fourier coefficients leads a contradiction.

It follows from Proposition 10.9 that if F is any modular form of degree 2 and even weight, then there is a polynomial $P(E_4, E_6, E_{12})$ such that $F - P$ vanishes on N. Now divide by Θ^2 and repeat the argument. At each stage the weight decreases by 10, so we eventually obtain a form of weight less than or equal to zero. Hence M_* is generated by the four modular forms E_4, E_6, E_{12}, and Θ^2.

The last generator can be replaced by E_{10}. We know that $E_{10} = \alpha E_4 E_6 + \beta \Theta^2$ for suitable complex constants α, β. Another comparison of Fourier coefficients shows that $\beta \neq 0$, proving Theorem 10.2.

THEOREM 10.2. *The graded ring M_* of modular forms of degree 2 and even weight is generated as an algebra over $\mathbb{C}$ by the four Eisenstein series of weight 4, 6, 10, and 12.*

We began Chap. 10 with a candidate for a homology theory, and we end it with another for coefficients of a cohomology theory, namely $\mathbb{Z}[1/30][E_4, E_6, E_{10}, E_{12}]$ (with perhaps E_{10} or some higher weight element inverted). If the pattern established by elliptic cohomology holds, then the first of these theories is a level 2 theory with coefficients in a ring where (1) 2 and 3 are invertible and (2) there are generators $x_8, x_{12}, x_{16}^{\pm 1}, \ldots$ corresponding to low-dimensional generators in $\Omega_*^{\langle 8 \rangle} \otimes \mathbb{Z}[1/6]$. Note: We must allow x_{16}^{-1}, given the identification of an $\mathbb{O}P^2$-bundle with its base.

The second theory will be a level one theory — hence the inversion of 5 as well as 2 and 3. If such a theory exists, we suspect that the associated genus bears a similar relation to the extended Witten genus $\beta^{\mathbb{O}}$, as the elliptic genus does to the extended Atiyah invariant $\beta^{\mathbb{H}}$. What makes this more than wishful thinking is the fact that [11] proves that $K3$-cohomology theories *do* exist; hence it is worthwhile to try to carry out the proposals just made in their framework. It is possible to be a little more precise: Starting with a generic Kummer surface in $\mathbb{C}P^3$, the associated formal group law (calculated using Stienstra's algorithm) corresponds to a homomorphism from $\Omega_*^{\langle 8 \rangle} \otimes \mathbb{Q}$ into $\mathbb{C}$. Prescribed generating manifolds map to complex images $x_{i_1}, x_{i_2}, \ldots$.

Rescaling the Jacobian $\mathrm{Jac}(C)$, which double covers the original surface, introduces modularity for $x_{i_1}, x_{i_2}, \ldots$, and varying defining parameters inside H_2 converts complex numbers into modular forms, defined at least on some subset of H_2. We end with one possibly significant example: In [58] Chapter VIII covers two specific defining equations. The first of these is

$$x^4 + y^4 + z^4 + t^4 + 2Dxyzt + A(x^2t^2 + y^2z^2) + \\ B(y^2t^2 + z^2x^2) + C(z^2t^2 + x^2y^2) = 0$$

with $D^2 + ABC - (A^2 + B^2 + C^2) + 4 = 0$ (see [58, Sec. 53]). Note: If $A = B = C = 0, D^2 + 4 = 0$, we obtain a particularly simple case. This choice of coordinates corresponds to the choice of a level structure on the 2-torsion subgroup (of order 16) in the covering Jacobian; it is suggestive given that with $O\langle 8 \rangle$-bordism as domain, we are concerned with abelian surfaces X with prescribed spin structure on the loop space LX. The cubic relation between A, B, C, and D is the equation for the moduli space of abelian surfaces with a level structure.

Hudson's second equation is

$$\begin{aligned}&\xi_4^2(u^2\xi_1^2+v^2\xi_2^2+w^2\xi_3^2-2vw\xi_2\xi_3-2wu\xi_3\xi_1-2uv\xi_1\xi_2)\\&\quad+2\xi_4\{vw\xi_1(\xi_2^2-\xi_3^2)+wu\xi_2(\xi_3^2-\xi_1^2)+uv\xi_3(\xi_1^2-\xi_2^2)+s\xi_1\xi_2\xi_3\}\\&\quad+(u\xi_2\xi_3+v\xi_3\xi_1+w\xi_1\xi_2)^2=0.\end{aligned}$$

Hudson refers to this as the equation referred to a Rosenhain tetrahedron (see [58, Sec. 54]). This corresponds to the choice of some nonisotropic affine subspace in the 2-torsion subgroup over $\mathbb{F}_2$, and the parameters u, v, w, s (the last of weight 2) are the moduli of such structures on abelian surfaces.

One contrast between these two equations is that the first is adapted to calculating the formal group law (by hand if $D^2+4=0$ and we replace the variable x by ix, using some computational package in general), while the second is equally adapted to the curve of genus 2 whose Jacobian is the double cover (compare Eq. 10.1 from [32] giving the locus of the Kummer surface).

10.5. Notes

This is a vast subject drawing on topology, algebraic geometry, (both pre- and post-Grothendieck) and complex analysis. We are fully aware of our limitations, which in part explains the tentative approach. Section 10.1 is a straightforward meditation on the work of Kreck and Stolz; for more on the Witten genus the reader is referred to [53, Chap. 6] and the calculations of [123].

Section 10.2 is the result of a crash course on abelian varieties, in which we found [74] with its numerous examples particularly useful. Notes by J. W. S. Cassels and E. V. Flynn [32] are harder to read but valuable because of the precise way the authors describe relations between geometric objects presented in terms of equations. They also provide an application of modern computational packages, which should be of immense value in testing some of the connections we suggested. Hudson's book on the Kummer surface [58] is quite simply a joy to read or dip into.

For the formal group law on $H^2(X, \mathbb{G}^{\wedge}_{m,X})$ with X a quartic surface embedded in $\mathbb{C}P^3$, we have found [103, 20] particularly useful. Jack Morava has suggested that Artin–Mazur conditions for the existence of a structure of this kind are almost identical with the standard folk definition of a low-dimensional Calabi–Yau manifold as an algebraic variety of dimension n such that $h^{i,0}=0$ for $0<i<n$ and $h^{n,0}=1$. This ties in with a suggestion made by E. Witten [121], that the Calabi–Yau setting is the correct one for a generalized elliptic cohomology. More recently Hopkins and Mahowald considered families of curves:

$$y^{p-1}=x^p+a_1x^{p-1}+\cdots+a_p$$

and their connection with the spectra EO_{p-1} generalizing EO_2. Again quoting Morava, the Jacobians of such curves — see the Lubin–Tate moduli space for height $p-1$ — there is a nice action of the group μ_{p-1} of roots of unity on such a curve, and the Jacobian decomposes into pieces, one of which is one-dimensional (generated by the elliptic differential dx/y). In comparison with these pyrotechnics our approach may seem pedestrian.

For Siegel modular forms we relied mainly on the published lectures of Klingen [65]. However for those with a knowledge of German and the wish to reach the proof of Theorem 10.2 as quickly as possible [45] is strongly recommended. One natural question is to ask what happens to the graded ring M_* when the full modular group $Sp_4(\mathbb{Z})$ is replaced by a discrete subgroup corresponding to $\Gamma_0(2) \subset Sp_2(\mathbb{Z})$. Proposition 10.9 suggests that (with the obvious notation) $\delta_4(z_1)\delta_4(z_4), \varepsilon_8(z_1)\varepsilon_8(z_4)$ and $\delta_4^4(z_1)\varepsilon_8^2(z_4) + \delta_4^4(z_4)\varepsilon_8^2(z_1)$ become the new generators of $M_*|N$, but what of Θ and Θ^2?

A

Brown–Peterson Cohomology

At various points in this book, notably Chap. 6, we used the properties of Brown–Peterson cohomology, a summand of complex cobordism localized at the prime p. In Appendix A we give some idea of how BP^* and its dual homology theory are constructed, and with some repetition, we pull together the properties used. In no way is our account a substitute for those in [1, 96].

Recall from Sec. 1.1 of Chap. 1 that the coefficient ring Ω^*_{SO} is generated over $\mathbb{Z}[1/2]$ by equivalence classes of manifolds distinguished by their Pontrjagin numbers. In the same way we can consider Ω^*_U the (co)bordism ring of stably complex manifolds, i.e., manifolds admitting a complex structure on their stable normal bundles, two such are equivalent if and only if they have the same Chern numbers. In the family of classifying spaces $\{MU_n : n \geqslant 0\}$ for this cohomology theory, MU_n is defined to be the Thom space (one-point compactification) of the total space of the universal n-plane bundle over the classifying space BU_n. The restriction from $(n+1)$ to n yields a Thom space $\Sigma^2 MU_n$ and hence a map of compactified total spaces $\Sigma^2 MU_n \to MU_{n+1}$. This sequence of spaces and maps illustrates the following general definition.

DEFINITION. A *spectrum* is a collection of spaces $\{E_n\}$ (defined for all large values of n) and maps $\Sigma E_n \to E_{n+1}$. The suspension spectrum of a space X is defined by $X_n = \Sigma^n X$, where each map is the identity. More generally $\{E_n\}$ is a *suspension spectrum* if $\Sigma E_n \simeq E_{n+1}$. The homotopy groups of E are defined by:

$$\pi_k(E) = \varinjlim_n \pi_{n+k}(E_n).$$

A map of spectra $f : E \to F$ is a collection of maps $E_n \underset{f_n}{\longrightarrow} \varinjlim_k \Omega^k F_{n+k}$, where Ω denotes base-point-preserving loops, and $f_n \simeq \Omega f_{n+1}$.

Note: When E and F are suspension spectra, since operations of suspension and taking loops are adjoint, a map of spectra is equivalent to a collection obtained by (de)suspending a single map $E_n \to F_n$ for some n.

DEFINITION. The *homotopy category of CW-spectra* contains objects that are spectra built from CW-complexes and morphisms that are homotopy classes of maps, as defined.

With this definition we can suppose without loss of generality that the structural map $\Sigma E_n \to E_{n+1}$ is actually an embedding of a subcomplex.

Much of the material in Chap. 6 and beyond concerns homology and cohomology groups associated with suitable spectra, thus

$$E_k(X) = \lim_{\longrightarrow n} \pi_{n+k}((X/A)_\wedge E_n)$$

with coefficients given by $\{\pi_k E\}$. The general method applies to both BP and its related p-local spectra and those introduced in Chap. 9 to extend the definition of elliptic homology from $\mathbb{Z}[1/2]$ to $\mathbb{Z}$.

One problem with the coefficient ring Ω_U^* is that having a generator in each even dimension, it is too large for comfort. However $\Omega_U^*(X)$ is an oriented cohomology theory, and $\Omega_U^*(\mathbb{C}P^\infty)$ is a polynomial ring over the coefficients. As explained in Chap. 2 taking, the first Chern class of the tensor product of two line bundles gives a formal group law whose logarithm can be written as:

$$\log y = \Sigma_n \frac{[\mathbb{C}P^n]}{n+1} y^{n+1}.$$

We next take cobordism with coefficients in a suitable subring R of the rational numbers $\mathbb{Q}$; the most important case is $R = \mathbb{Z}_{(p)}$, the ring of rational numbers with denominators prime to p. The existence of the p-local spectrum BP follows from Theorem A.1.

THEOREM A.1. *Let $d > 1$ be a natural number and $R \subseteq \mathbb{Q}$ some subring containing d^{-1}. There is a unique map of spectra $e = e_d : MU_R \to MU_R$ such that:*

1. *$e^2 = e$ (e is idempotent).*
2. *On $\pi_*(MU_R)$ e acts by:*

$$e[\mathbb{C}P^n] = \begin{cases} 0, & d \mid n+1 \\ [\mathbb{C}P^n], & d \nmid n+1 \end{cases},$$

3. *The maps e_{d_1} and e_{d_2} commute.*

COROLLARY A.1. *For each prime number p there is a unique map of spectra $\varepsilon = \varepsilon_p : MU_{\mathbb{Z}_{(p)}} \to MU_{\mathbb{Z}_{(p)}}$ such that:*

1. $\varepsilon = \varepsilon^2$ *(ε is idempotent).*

2. $\varepsilon[\mathbb{C}P^n] = \begin{cases} [\mathbb{C}P^n], & n = p^t - 1 \\ 0, & \textit{otherwise} \end{cases}.$

Proof: Corollary A.1 follows from the theorem by localizing at all primes $q \neq p$. We prove Theorem A.1 by modifying the definition of the logarithm (known to be equivalent to the formal group law) and deducing the existence of the map e_d from the universality of the formal group law for MU. The form of the modified logarithm explains why d is needed as a denominator. Write

$$\operatorname{mog} y = \log y - \frac{1}{d}(\log \zeta_1 y + \cdots + \log \zeta_d y),$$

where $\zeta_1, \ldots, \zeta_d$ are the dth roots of unity, that is, the d roots of the equation $y^d - 1 = 0$. Coefficients of $\operatorname{mog} y$ lie in $\pi_*(MU) \otimes \mathbb{Q}[e^{2\pi i/d}]$; if μ denotes the formal group law:

$$x_1 +_\mu x_2 = \exp(\operatorname{mog} x_1 + \operatorname{mog} x_2).$$

The expression $\zeta_1 y +_\mu \cdots +_\mu \zeta_d y$ is a formal power series with coefficients in $\pi_*(MU) \otimes \mathbb{Z}[\zeta_1, \ldots, \zeta_d]$. Symmetry in the roots ζ_j means that these coefficients can be expressed in terms of elementary symmetric functions $\sigma_j (1 \leqslant j \leqslant d)$. The equation $y^d - 1 = 0$ shows that $\sigma_1 = \cdots = \sigma_{d-1} = 0$ and $\sigma_d = (-1)^{d-1}$, and Theorem A.1, Part 2 follows by interpreting this for the function mog.

If $e^2 = e$ is a map on homotopy groups, then $e^2 = e$ in the category of CW-spectra. A similar argument shows that maps e_{d_1} and e_{d_2} commute. ■

We showed that $\pi_* BP \cong \mathbb{Z}_{(p)}[v_1, v_2, \ldots]$, with $\dim v_i = 2p^i - 2$. Let I_n be the ideal generated by $p, v_1, \ldots, v_{n-1}$. Consider the rings:

$$\begin{aligned} BP\langle n\rangle_* &= \mathbb{Z}_{(p)}[v_1, \ldots, v_n] \\ P(n)_* &= BP_*/I_n \\ E(n)_* &= \mathbb{Z}_{(p)}[v_1, \ldots, v_n, v_n^{-1}] \\ B(n)_* &= v_n^{-1} BP_*/I_n, \\ K(n)_* &= \mathbb{F}_p[v_n, v_n^{-1}]. \end{aligned}$$

THEOREM A.2 (LANDWEBER EXACTNESS). *The functor $X \mapsto BP_*(X) \underset{BP_*}{\otimes} M$ for the fixed BP_*-module M is a generalized homology theory if and only if for each prime p and natural number n multiplication by v_n in $M \underset{BP_*}{\otimes} (BP_*/I_n)$ is injective.*

Granted this result (the proof is outlined in [97, p. 171–173]), it follows easily that $E(n)_*$ corresponds to a homology theory. Note: For $i > n$ $E(n)_* \bigotimes_{BP_*} (BP_*/I_i) = 0$.

For small values of n note that $E(1) =$ a summand of periodic complex K-theory localized at p and $E(2)$ is similarly related to elliptic cohomology. The emerging pattern in Chaps. 9 and 10 suggests that $E(4)$ splits off from the spectrum of the conjectural theory $\pi E_*^{\mathbb{O}}$, again localized at $p(\geqslant 5)$.

The remaining theories are not quotients of BP_* in the sense of the Landweber theorem, although they are generalized homology theories. We indicated earlier how to show this using the Baas–Sullivan construction; other methods of proof are available. See for example the survey article by May [85] and its bibliography. In the case of the Morava K-theories the isomorphism:

$$K(n)_*(BG) = BP_*(BG) \bigotimes_{BP_*} K(n)_*$$

holds only for a restricted class of p-groups G.

The central role of $K(n)_*$ can be illustrated as follows: In one direction after suitable completion, the spectrum $B(n)$ splits into a wedge of suspensions of $K(n)$. We can be more precise: Chapter 6 results should be seen in the light of recent work by Ravenel et al. [98]. Let $P(n)^*$ be the cohomology defined by the spectrum $P(n)$ except that $P(0)^*(X)$ equals the p-adic completion of $BP^*(X)$. In the special case when $X = BG$, we can take $P(0)^*(BG) = BP^*(BG)$, as explained long ago in the case of K-theory by Atiyah [13]. Among other results we have

- If $K(n)^*(X)$ is concentrated in even dimensions for infinitely many values of n, then it is concentrated in even dimensions for all values of n.

- If $K(n)^{\mathrm{odd}}(X) = 0$ for all $n > 0$, then $P(k)^{\mathrm{odd}}(X) = 0$ for all values of $k \geqslant 0$, and $P(k)^{\mathrm{even}}(X)$ is a flat $P(k)^*$-module in the category of $P(k)^*(P(k))$-modules which are finitely presented over $P(k)^*$.

As a special case we see that if G is a finite p-group such that $K(n)^{\mathrm{odd}}(BG) = 0$ for all $n > 0$, then $BP^{\mathrm{odd}}(BG) = 0$; since each $E(n)^*$ is a quotient theory of BP^*, $E(n)^{\mathrm{odd}}(BG) = 0$ also.

For more information we refer the reader to [96, Chap. 4], [97, pp. 166–178], and [1, Sec. 15].

B

Cayley Projective Plane $\mathbb{O}P^2$

We list here some of the main properties of $\mathbb{O}P^2$. The existence of this space is something of an anomaly, and unlike projective spaces defined over $\mathbb{R}$, $\mathbb{C}$, and $\mathbb{H}$, it does not begin an infinite sequence. We start with Theorem B.1.

THEOREM B.1. *There is a unique normed algebra of dimension 8 over* $\mathbb{R}$.

In a normed algebra $\|xy\| = \|x\|\,\|y\|$. We do not assume that our algebra is associative or commutative. The main step in the proof of both existence and uniqueness is to show that $\mathbb{O}$ contains a subalgebra $\mathbb{H}$ with the usual generators $\{1,i,j,ij\}$ and then that $\mathbb{O} \cong \mathbb{H}\oplus\mathbb{H}$, with the product given by:

$$(a,b)(c,d) = (ac-\bar{d}b, da+b\bar{c})$$

whose conjugation is quaternionic. Over $\mathbb{O}$ itself $\overline{(a,b)} = (\bar{a},-b)$. The $+1$ eigenspace is generated by $(1,0)$, and it is one-dimensional over $\mathbb{R}$. The -1 eigenspace is seven-dimensional, and it contains the pure imaginary Cayley numbers. The basic properties of $\mathbb{O}$ are

1. $\bar{\bar{x}} = x$, $x\bar{x} = \|x\|^2\,1 = \bar{x}x$, $\overline{xy} = \bar{y}\bar{x}$.

2. The real parts of xy and yx are equal, and for pure imaginary numbers, $\mathrm{Im}(xy) = \mathrm{Im}(yx)$.

3. The inner product of x and y equals that of $\bar{x}$ and $\bar{y}$; the common value is $\mathrm{Re}(x\bar{y}) = \mathrm{Re}(\bar{y}x)$.

4. If $[x,y,z] = (xy)z - x(yz)$ denotes the *associator* of $x,y,z \in \mathbb{O}$, then $[\ ,\ ,\ ]$ is an alternating function. These properties can be proved by direct calculation. As a consequence we have Property 5.

5. $\|xy\| = \|x\|\,\|y\|$. This follows from the sequence of equalities:

$$\|xy\|^2 = \mathrm{Re}((\overline{xy})(xy)) = \mathrm{Re}(\bar{y}\bar{x}(xy)) = \mathrm{Re}(\bar{y}(\bar{x}(xy)))$$

by the limited associativity implied by the alternating property in 4

$$= \mathrm{Re}(\bar{y}(\bar{x}x)y) \text{ (direct calculation)} = \|x\|^2\,\mathrm{Re}(\bar{y}y) = \|x\|^2\,\|y\|^2.$$

The algebra $\mathbb{O}$ is closely related to the exceptional Lie groups; it can be shown for example that $\mathrm{Aut}_{\mathbb{R}}(\mathbb{O}) \cong G_2$. This follows by examining stabilizers of preferred elements of $\mathbb{O}$ under the action of the automorphism group, then comparing these subgroups with the known subgroup structure of G_2.

To construct the projective plane $\mathbb{O}P^2$ explicitly, we need the *exceptional Jordan algebra*

$$J = \{3\times 3 \text{ Hermitian matrices over } \mathbb{O}, A = \begin{pmatrix} \lambda_1 & x_3 & \bar{x}_2 \\ \bar{x}_3 & \lambda_2 & x_1 \\ x_2 & \bar{x}_1 & \lambda_3 \end{pmatrix}, \quad \lambda_i \in \mathbb{R}\}.$$

The geometry of the projective plane $KP^2 (K = \mathbb{R}, \mathbb{C}$ or $\mathbb{H})$ depends on being able to do linear algebra over the field K, i.e., manipulate relations between points described in terms of homogeneous coordinates. As a substitute for this over $\mathbb{O}$, we use structural maps of J, derived from its product $[A\circ B = (AB+BA)/2]$ and $\ell(A) = \lambda_1+\lambda_2+\lambda_3$. Define a bilinear map by $b(A,B) = \ell(A\circ B)$ and a trilinear map by $t(A,B,C) = b(A\circ B, C)$. We can then prove Theorem B.2.

THEOREM B.2.

1. *F_4 is isomorphic to the group of product-preserving automorphisms of the algebra J.*

2a. *The isotropy subgroup of* $e_1 = \mathrm{diag}(1,0,0)$ *is isomorphic to Spin* (9).

2b. *The isotropy subgroup of the triple of diagonal matrices* $\{e_1,e_2,e_3\}$ *is isomorphic to Spin* (8).

REMARKS. In proving Part 2b it is easy to see (for example) that fixing e_1 implies invariance of the space of matrices:

$$V_2+V_3 = \left\{\begin{pmatrix} 0 & x_3 & \bar{x}_2 \\ \bar{x}_3 & 0 & x_1 \\ x_2 & \bar{x}_1 & 0 \end{pmatrix} : x_i \in \mathbb{O}\right\}.$$

Hence fixing e_1, e_2, and e_3 separately implies the invariance of each of the eight-dimensional spaces V_i. For Part 2a, fixing e_1 implies only that e_2+e_3 is mapped to some $\begin{pmatrix} 0 & 0 & 0 \\ 0 & \lambda & x_1 \\ 0 & \bar{x}_1 & -\lambda \end{pmatrix}$, with $\lambda^2 + \bar{x}_1 x_1 = 2$. Part 1 requires knowledge of the 26-dimensional representation of F_4 (corresponding to the subspace of J on which $\ell = 0$) to show that F_4 maps injectively into Aut (J). Surjectivity depends on showing that up to the F_4-action, an element of J is determined by $\ell(A)$, $b(A,A)$, and $t(A,A,A)$.

We are now ready for the Cayley plane itself. A point in KP^2 (K equals any field of coefficients) is a one-dimensional subspace of K^3; hence it can be identified with the orthogonal projection operator onto this subspace. This has rank equal to 1, and it is idempotent and Hermitian. With $K = \mathbb{O}$, consider the F_4 orbit of $e_1 \in J$, which can be shown to coincide with $\{A : A \in J, \ell(A) = b(A,A) = t(A,A,A) = 1\}$. Furthermore such matrices have the three properties of the projection operators already mentioned.

DEFINITION. $\mathbb{O}P^2 = F_4/\mathrm{Spin}(9)$.

As a diagonal matrix operating on J via the product $\circ$, e_1 has a 0-eigenspace of dimension 10 generated by e_2, e_3, and V_1, while $V_2 + V_3$ defines the tangent space to $\mathbb{O}P^2$ at e_1. The points of $\mathbb{O}P^2$ in the 0-eigenspace, i.e., rank one idempotents orthogonal to e_1, describe an eight-dimensional sphere carrying a Spin (9)-action, and as such these form a Cayley projective line. Collinearity is decided by a determinantal condition, which is suggested by the constant term in the Cayley–Hamilton equation for an element A of J:

$$A^3 - c_1 A^2 + c_2 A - c_3 1_3 = 0.$$

Here

$$\begin{aligned} c_1 &= \ell(A), \\ c_2 &= \frac{1}{2}(\ell(A)^2 - b(A,A)), \\ c_3 &= \frac{1}{6}(\ell(A)^3 - 3\ell(A)b(A,A) + 2t(A,A,A)). \end{aligned}$$

We state the necessary and sufficient condition for the points A, B, and C to be collinear as

$$\ell(A)\ell(B)\ell(C) - (\ell(A)b(B,C) + \ell(B)b(C,A) + \ell(C)b(A,B)) + 2t(A,B,C) = 0.$$

Following this line of thought, we can show that Desargues's theorem on triangles in perspective fails in $\mathbb{O}P^2$. This in itself suffices to show that there can be no higher dimensional projective spaces, since if these were to exist Desargues's theorem would follow from the incidence axioms of projective geometry.

For the details of the argument just outlined, see [4, Secs. 17–18]. The topological properties of $\mathbb{O}P^2$ are given in [24], and these can be summarized as follows: There is a generator u of $H^8(\mathbb{O}P^2, \mathbb{Z})$ such that the integral Pontrjagin classes are given by $p_2 = 6u$ and $p_4 = 39u^2$. Hence with $\mathbb{O}P^2$ oriented by the class dual to u^2, the nonvanishing Pontrjagin numbers are 36 and 39. The signature equals 1, and $\hat{A}$-genus equals 0. These tangential invariants are such that $\mathbb{O}P^2$ admits no almost-complex structure associated with the usual C^∞-structure.

C

Index of $\Gamma_0(N)$ in $SL_2(\mathbb{Z})$

The principal congruence subgroup $\Gamma(N)$ of level N in $SL_2(\mathbb{Z})$ is defined to be the kernel of the natural projection homomorphism from $SL_2(\mathbb{Z})$ onto $SL_2(\mathbb{Z}/N)$; thus:

$$\Gamma(N) := \left\{A \in SL_2(\mathbb{Z}) : A \equiv \left(\begin{smallmatrix} 1 & 0 \\ 0 & 1 \end{smallmatrix}\right) \pmod{N}\right\}.$$

There are inclusions $\Gamma_0(N) \subset \Gamma_1(N) \subset \Gamma(N)$, where:

$$\Gamma_0(N) := \left\{\left(\begin{smallmatrix} a & b \\ c & d \end{smallmatrix}\right) \in SL_2(\mathbb{Z}) : c \equiv 0 \pmod{N}\right\},$$
$$\Gamma_1(N) := \left\{\left(\begin{smallmatrix} a & b \\ c & d \end{smallmatrix}\right) \in SL_2(\mathbb{Z}) : a,d \equiv 1 \pmod{N} \text{ and } c \equiv 0 \pmod{N}\right\}.$$

PROPOSITION C.1. *The indices of* $\Gamma(N)$, $\Gamma_1(N)$, *and* $\Gamma_0(N)$ *in* $SL_2(\mathbb{Z})$ *are respectively* $N^3\prod_{p|N}(1-1/p^2)$, $N^2\prod_{p|N}(1-1/p^2)$, *and* $N\prod_{p|N}(1+1/p)$.

Proof: For the index of $\Gamma(N)$, see [95, pp. 20–23]; the two other cases are similar. The argument proceeds by counting primitive pairs $(c,d) \pmod{N}$, i.e., pairs of coprime integers, where both are also coprime with N. Each such pair is realized in $SL_2(\mathbb{Z})$ by N matrices $\pmod{N}$. If $\lambda(N)$ equals the number of incongruent primitive pairs, $\lambda(N)$ is a multiplicative function. Explicit counting shows that if p is prime and $t \geqslant 1$, $\lambda(p^t) = p^{2t}(1-1/p^2)$, which implies the result. ∎

In Chap. 4 we saw that the index of $\Gamma_0(N)$ plays a part in calculating pairs ($N =$ level, $k =$ weight) of integers for which we obtain one-dimensional eigenspaces of modular forms with respect to the action of the Hecke algebra. This involved solving the equation

$$\frac{1}{12}([SL_2(\mathbb{Z}),\Gamma_0(N)]k) = \sum_{d|N} \varphi\left(\left(d, \frac{N}{d}\right)\right),$$

Table C.1. Symplectic Fixed Points

$n = \|g\|$	1	2	3	4	5	6	7	8	9	11	12	15	16	23
$\chi(g)$	24	8	6	4	4	2	3	2	2	2	1	1	1	1

where φ is the Euler phi function. It turned out that all but two of the generating cusp forms (constructed by using the Dedekind η-function) corresponded to even conjugacy classes of elements in the Mathieu group M_{24}. Like much summarized under the heading moonshine, this phenomenon has yet to receive a satisfactory explanation. What is equally amazing is that both the index of $\Gamma_0(N)$ and the Mathieu groups appear in the theory of $K3$-surfaces, for which, as explained in Chap. 10, the formal Brauer group provides a good one-dimensional formal group law. This we now explain; the two references are [89, 82]. Let X be a $K3$-surface with nonvanishing holomorphic 2-form ω and G the finite group of automorphisms of X that preserve ω. We already saw that the rational vector space $H^*(X,\mathbb{Q})$ has dimension 24. Considering the induced G-action, the fact that ω is preserved implies that G acts trivially on $H^{2,0}$ and hence also on $H^{0,2}$. Clearly the action is trivial on H^0 and H^4, and [89] shows that $H^{1,1}$ also contains a nontrivial G-invariant element:

$$\dim_{\mathbb{Q}}(H^*(X,\mathbb{Q}))^G \geqslant 5.$$

Using the Atiyah–Singer index theorem and the Lefschetz fixed-point formula, Mukai next shows, that as a rational representation space, $H^*(X,\mathbb{Q})$ has character given by:

$$\chi(g) = \frac{24}{[SL_2(\mathbb{Z}),\Gamma_0(n)]} = \frac{24}{n}\left(\prod_{p|n}\left(1+\frac{1}{p}\right)\right)^{-1},$$

where n is the order of the element g. This gives Table C.1. Comparison with the usual permutation character of M_{23} reveals a remarkable coincidence of values. This is made explicit in the following result, where the dimensional bound on the fixed-point subset is rephrased as G has at least five orbits in the 24 symbols being permuted.

THEOREM C.1. *If G is a group of ω-preserving automorphisms of the $K3$-surface X, then there is an embedding $i : G \to M_{23}$ such that $i(G)$ is conjugate to a subgroup of one of 11 maximal subgroups of M_{23} satisfying the five-orbit condition.*

COROLLARY C.1. *The embedding i induces an isomorphism between the representation space $H^*(X,\mathbb{Q})$ and the restriction of the natural permutation representation of M_{23} to G.*

Table C.2. Examples of Automorphism Groups

Group	Order	Surface
$L_2(7)$	168	$x^3y+y^3z+z^3x+t^4=0$
$M_{20}\cong(2^4):A_5$	960	$x^4+y^4+z^4+t^4+12xyzt=0$
$(4^2):S_4$	384	$x^4+y^4+z^4+t^4=0$
$(Q_8*Q_8):S_3$	192	$x^4+\cdots+t^4-\sqrt{-12}(x^2y^2+z^2t^2)=0$

The reader is warned that there are slight discrepancies between the lists of 11 groups given in [89] and [82]. Table C.1 shows Mukai's examples of $K3$-surfaces that are embedded in $\mathbb{C}P^3$ and their full ω-preserving automorphism groups.

Our notation is standard; a colon (:) stands for a semidirect product, where the first group is normal; $Q_8 * Q_8$ denotes the central product of two copies of the quaternion group.

REMARKS. The first of these examples goes back to F. Klein, who identified $L_2(7)$ with the complex automorphism group of the genus 3 curve defined by $x^3y+y^3z+z^3x=0$. For the third, combine permutation of the coordinates with automorphisms of the type:

$$(x:y:z:t)\mapsto(i^ax:i^by:i^cz:i^dt)\qquad i=\sqrt{-1}.$$

Is it significant that a large automorphism group is associated with a curve of the second type already mentioned in connection with a choice of level 2 structure in Notes to Chap. 10?

The main steps in Mason's proof of Theorem C.1 are

1. A 2-Sylow subgroup of G is isomorphic to a subgroup of M_{23}.
2. The order of G divides $2^7\cdot3^2\cdot5\cdot7$.
3. If 7 divides the order of G, G is isomorphic to a subgroup of $L_2(7)$.
4. If 5 divides the order of G, 7 does not by Step 3 and a theorem of R. Brauer describes the nonsolvable examples. Otherwise we can effectively construct a composition series for a maximal example.
5. We must do a case-by-case analysis of the possible groups, all of which are solvable, of order 2^a3^b with $a\leqslant7$ and $b\leqslant2$.

References

1. J. F. Adams, *Stable Homotopy and Generalized Homology*, Univ. of Chicago Press, Chicago (1974).
2. J. F. Adams, "Maps between classifying spaces II," *Invent. Math.* **49**, pp. 1–65 (1978).
3. J. F. Adams, "Graeme Segal's Burnside ring conjecture," *Bull. Amer. Math. Soc.* **6**, 201–210 (1982).
4. J. F. Adams, *Lectures on Exceptional Lie Groups*, Univ. of Chicago Press, Chicago (1996).
5. J. F. Adams, J. P. Haeberly, S. Jackowski, and J. P. May, "A generalization of the Atiyah–Segal completion theorem," *Topology* **27**(1), 1–6 (1988).
6. J. F. Adams, J. P. Haeberly, S. Jackowski, and J. P. May, "A generalization of the Segal conjecture," *Topology* **27**(1), 7–21 (1988).
7. A. Adem, "Characters and K-theory of discrete groups," *Invent. Math.* **114**, 489–514 (1993).
8. A. Adem and R. J. Milgram, *Cohomology of Finite Groups*, Grundlehren der mathematischen Wissenschaften 209, Springer, Heidelberg (1994).
9. G. E. Andrews, R. A. Askey, B. C. Berndt, K. G. Ramanathan, and R. A. Rankin, *Ramanujan Revisited*, (Proceedings of the Centenary Conference), Academic Press, New York (1988).
10. M. Artin, "Supersingular $K3$ surfaces," *Ann. Sci. ENS* **7**, 165–189 (1974).
11. M. Artin and B. Mazur, "Formal groups arising from algebraic varieties," *Ann. Sci. ENS* **10**, 87–131 (1997).
12. M. Aschbacher, *Sporadic Groups*, Cambridge studies in advanced mathematics, 104, Cambridge, Univ. Press, Cambridge, UK (1994).
13. M. F. Atiyah, "Characters and cohomology of finite groups," *Publ. Math. IHES* **9**, 247–289 (1961).
14. M. F. Atiyah, "Topological quantum field theories," *Publ. Math. IHES* **68**, 175–186 (1989).
15. M. F. Atiyah, R. Bott, and A. Shapiro, "Clifford modules," *Topology 3, suppl.* **1**, 3–38 (1964).
16. M. F. Atiyah and G. B. Segal, "Equivariant K-theory and completion," *J. Diff. Geom.* **3**, 1–18 (1969).
17. N. Baas, "On bordism theories of manifolds with singularities," *Math. Scan.* **33**, 270–302 (1973).
18. A. Baker, "Elliptic genera of level N and elliptic cohomology," *J. London Math. Soc. (2)* **49**, 581–594 (1994).

19. A. Baker and C. Thomas, *Classifying Spaces, Virasoro Equivariant Bundles, Elliptic Cohomology and Monshine*, ETH preprint (1996).

20. F. Beukers and J. Stienstra, "On the Picard–Fuchs equation and the formal Brauer group of certain elliptic $K3$ surfaces," *Math. Annalen* **271**, 269–304 (1985).

21. B. Birch and W. Knyk, *Modular Functions of One Variable IV*, (Antwerp), Lecture Notes in Mathematics 476, Springer (1975).

22. N. Blackburn, "Generalizations of certain elementary theorems on p-groups," *Proc. London Math. Soc.* **11**, 1–22 (1961).

23. R. E. Borcherds, "Monstrous moonshine and monstrous Lie superalgebras," *Invent. Math.* **109**, 405–444 (1992).

24. A. Borel and F. Hirzebruch, "Characteristic classes and homogeneous spaces, I," *Amer. J. Math.* **80**, 458–538 (1958); "Characteristic classes and homogeneous spaces, II," *Amer. J. Math.* **81**, 315–382 (1959).

25. R. Bott and C. H. Taubes, "On the rigidity theorems of Witten," *J. Amer. Math. Soc.* **2**, 137–186 (1989).

26. L. Breen, *Fonctions Théta et le Théorème du Cube*, Lecture Notes in Mathematics 980, Springer (1983).

27. E. H. Brown and F. P. Peterson, "A spectrum whose $\mathbb{Z}_p$ cohomology is the algebra of reduced pth powers," *Topology* **5**, 149–154 (1966).

28. M. Brunetti, "A family of $2(p-1)$-sparse cohomology theories and some actions on $h^*(BC_{p^n})$," *Math. Proc. Camb. Phil. Soc.* **116**, 223–228 (1994).

29. J. L. Brylinski, "Representations of loops groups, Dirac operators in loop space and modular forms," *Topology* **29**, 461–480 (1990).

30. W. Burnside, *Theory of Groups of Finite Order*, Dover, NY (1955).

31. G. Carlsson, "Equivariant stable homotopy theory and Segal's Burnside ring conjecture," *Annals of Math.* (2) **20**, 189–224 (1984).

32. J. W. S. Cassels and E. V. Flynn, *Prolegomena to a Middlebrow Arithmetic of Curves of Genus 2*, LMS Lecture Notes 230, Cambridge Univ. Press (1996).

33. P. E. Conner and E. E. Floyd, *Differentiable Periodic Maps*, (2nd ed.), Lecture Notes in Mathematics 738, Springer (1979).

34. P. E. Conner and E. E. Floyd, *The Relation of Cobordism to K-theories*, Lecture Notes in Mathematics 28, Springer (1996).

35. J. Conway, R. T. Curtis, S. P. Norton, R. A. Parker, and R. A. Wilson, *An Atlas of Finite Groups*, Clarendon Press, Oxford, UK (1985).

36. J. Conway and S. P. Norton, "Monstrous Moonshine," *Bull. London Math. Soc.* **11**, 308–330 (1979).

37. P. Deligne and M. Rapoport, *Les Schémas de Modules de Courbes Elliptiques*, Lecture Notes in Mathematics 349, Springer, pp. 143–316 (1972).

38. J. Devoto, *An Algebraic Description of the Elliptic Cohomology of Classifying Spaces*, (submitted).

39. J. Devoto, "Equivariant elliptic cohomology and finite groups," *Mich. Math. J.* **43**, 3–32 (1996).

40. T. tom Dieck, *Transformation Groups and Representation Theory*, Lecture Notes in Mathematics 766, Springer (1979).

41. A. Dold, Halbexakte Homotopiefunktoren, Lecture Notes in Mathematics 12, Springer (1966).

42. C. Dong and G. Mason, "An orbifold theory of genus zero associated to the sporadic group M_{24}," *Comm. Math. Physics* **164**, 87–104 (1994).

43. E. Dyer, *Cohomology Theories*, Benjamin, New York (1969).

44. L. Euler, De integratione aequationis differentialis $(m\,dx/\sqrt{1-x^4}) - (n\,dy/\sqrt{1-y^4})$, Commentatio 251 indictis Enestroemiani, *Novi Commentarii Academiae Scientiarum Petropolitanae* **6**, pp. 37–57, Summarium ibidem, pp. 7–9 (1756/7, 1761). Or *Opera Omnia, Series I, Opera Mathematica, Volumen XX, Commentationes Analyticae ad Theoriam Integralium Ellipticorum Pertinentes, Volumen Prius*, ed. Adolf Krazer, Lipsiae et Berolini, B. G. Teubner, MCMXII, reprint, pp. 58–79 (1978).

45. E. Freitag, *Siegelsche Modulfunktionen*, Die Grundlehren der mathematischen Wissenschaften 254, Springer, Heidelberg (1983).

46. I. Frenkel, J. Lepowsky, and A. Meurman, *Vertex Operator Algebras and the Monster*, Academic, Boston (1988).

47. V. Giambalvo, "$\langle 8\rangle$-cobordism," *Ill. J. Math.* **15**, 533–541 (1971).

48. V. Ginzburg, M. Kapranov, and E. Vasserot, *Elliptic Algebras and Equivariant Elliptic Cohomology*, Univ. of Chicago preprint (1995).

49. D. J. Green, "The 3-local cohomology of the Matthieu group M_{24}," *Glasgow Math. J.* **38**, 69–75 (1996).

50. P. Griffiths and J. Harris, *Principles of Algebraic Geometry*, J. Wiley, New York (1978).

51. B. Gross, Letter to P. Landweber dated 7 April 1986.

52. M. Hazewinkel, *Formal Groups and Applications*, Academic, New York (1978).

53. F. Hirzebruch, T. Berger, and R. Jung, *Manifolds and Modular Forms* Vieweg, Braunschweig/Wiesbaden, Germany (1992).

54. N. Hitchin, "Harmonic spinors," *Advances in Math.* **14**, 1–55 (197).

55. M. Hopkins, *Topological Modular Forms, the Witten Genus and the Theorem of the Cube*, ICM Proceedings, Birkhäuser Zürich (1994).

56. M. Hopkins, N. Kuhn, and D. Ravenel, *Generalized Group Characters and Complex Oriented Cohomology Theories*, (submitted).

57. M. Hopkins, N. Kuhn, and D. Ravenel, *Morava K-theories of Classifying Spaces and Generalized Characters for Finite Groups*, Lecture Notes in Mathematics 1509, Springer, pp. 186–209 (1992).

58. R. W. H. T. Hudson, *Kummer's Quartic Surfaces*, (reprint of 1905 ed.), Cambridge Univ. Press, Cambridge, UK (1990).

59. B. Huppert, *Endliche Gruppen I*, Grundlehren der mathematischen Wissenshaften 134, Springer, Heidelberg (1967).

60. D. Husemoller, *Fibre Bundles*, Springer, New York (1974).

61. I. Igusa, "On the transformation theory of elliptic functions," *Amer. J. Math.* **81**, 436–452 (1959).

62. G. D. James, "Modular characters of the Mathieu groups," *J. of Algebra* **27**, 57–111 (1973).

63. C. Jansen, K. Lax, R. Parker, and R. Wilson, *An Atlas of Brauer Characters*, Clarendon Press, Oxford, UK (1995).

64. N. Katz, *p-adic Properties of Modular Schemes and Modular Forms*, Lecture Notes in Mathematics 350, Springer, pp. 69–191 (1973).

65. H. Klingen, *Introductory Lectures on Siegel Modular Forms*, Cambridge Studies in Advanced Mathematics, Cambridge, UK (1990).

66. A. W. Knapp, *Elliptic Curves*, (Math. Notes 40), Princeton Univ. Press, Princeton (1992).

67. M. Kreck and S. Stolz, "$\mathbb{H}P^2$-bundles and elliptic homology," *Acta Math.* **171**, 232–261 (1993).

68. I. Kriz, "Morava K-theory of classifying spaces — some calculations," *Topology* **37**, in press (1998).

69. P. S. Landweber, *Elliptic Cohomology and Modular Forms*, Lecture Notes in Mathematics 1326, Springer, pp. 55–68 (1988).

70. P. S. Landweber, *Supersingular Elliptic Curves and Congruences for Legendre Polynomials*, Lecture Notes in Mathematics 1326, Springer, pp. 68–93 (1988).

71. S. Lang, *Differential Manifolds*, Addison Wesley, Reading, MA (1972).

72. S. Lang, *Cyclotomic Fields*, Graduate texts in Math. 59, Springer, Heidelberg (1978).

73. S. Lang, *Elliptic Functions*, Springer, New York (1987).

74. H. Lange and Ch. Birkenhake, *Complex Abelian Varieties*, Grundlehren der mathematischen Wissenschaften 302, Springer, Heidelberg (1992).

75. K. Lee and I. Kriz, "Odd degree elements in the Morava $K(n)$-cohomology of finite groups," *Topology and Its Applications* (to appear).

76. G. Lewis, "The integral cohomology rings of groups of order p^3," *Trans. Ameri. Math. Soc.* **132**, 501–529 (1968).

77. L. Lewis Jr, J. P. May, and M. Steinberger, *Equivariant Stable Homotopy Theory*, Lecture Notes in Mathematics 1213, Springer (1980).

78. I. Madsen, *Smooth Spherical Space Forms*, Lecture Notes in Mathematics 657, Springer, pp. 303–352 (1978).

79. Y. Marttin and K. Ono, "Eta quotients and Elliptic Curves," *Proc. Amer. Math. Soc.* **125**, 3169–3176 (1997).

80. G. Mason, "Elliptic systems and the eta-function," *Notas Soc. Mat. Chile* **8**, 37–53 (1984).

81. G. Mason, "M_{24} and certain automorphic forms," *Contemporary Math.* **45**, 223–244 (1985).

82. G. Mason, "Symplectic automorphisms of $K3$-surfaces," *Centrum voor Wiskunde en Informatica Newsletter* **13**, 3–19 (1986).

83. G. Mason, "Finite groups and modular functions," *Proc. of Symposia in Pure Math.* **47**, 181–210 (1987).

84. G. Mason, "On a system of elliptic modular forms associated to the large Mathieu group," *Nagoya Math. J.* **118**, 177–193 (1990).

85. J. P. May, "Stable algebraic topology and stable topological algebra," *Bull. London Math. Soc.* **30**, 225–234 (1998).

86. J. Milnor and J. B. Stasheff, *Characteristic Classes*, Annals of Math. Studies 76, Princeton Univ. Press, Princeton (1974).

87. O. K. Mironov, "Multiplication in cobordism theories with singularities and Steenrod–tom Dieck operations," *Math. USSR Izvestija* **13**, 80–106 (1978).

88. S. A. Mitchell, *On the Lichtenbaum–Quillen Conjectures from a Stable Homotopy Theoretic Viewpoint*, Algebraic topology and its applications, MSRI27, Springer, Heidelberg, pp. 163–267 (1993).

89. S. Mukai, "Finite groups of automorphisms of $K3$-surfaces and the Mathieu group," *Invent. math.* **94**, 183–221 (1998).

90. S. Ochanine, "Sur les genres multiplicatifs définis par des intégrales elliptiques," *Topology* **26**, 143–151 (1987).

91. S. Ochanine, "Elliptic genera, modular forms over KO_* and the Brown–Kervaire invariant," *Math. Zeitschrift* **206**, 277–291 (1991).

92. A. Pressley and G. Segal, *Loop Groups*, Oxford Univ. Press, Oxford, UK (1986).

93. D. Quillen, "The spectrum of an equivariant cohomology ring I," *Annals of Math (2)* **94**, 549–572 (1971).

94. F. Quinn, "Lectures on axiometic TQFT," *IAS Park City Math. Series* **1**, 325 *et seq.* (1995).

95. R. A. Rankin, *Modular Forms and Functions*, Cambridge Univ. Press, Cambridge, UK (1977).

96. D. C. Ravenel, *Complex Cobordism and Stable Homotopy Groups of Spheres*, Academic Press, New York (1986).

97. D. C. Ravenel, *Nilpotence and Periodicity in Stable Homotopy Theory*, Annals of Math. Studies 128, Princeton Univ. Press, Princeton (1992).

98. D. Ravenel, S. Wilson, and T. Yagita, *Brown–Peterson Cohomology from Morava K-Theory*, (submitted).

99. D. Rector, "Modular characters and K-theory with coefficients in a finite field," *J. Pure and Applied Alg.* **4**, 137–158 (1974).

100. M. Rosen, "Abelian Varieties over $\mathbb{C}$," in: *Arithmetic Geometry*, Cornell and J. Silverman (eds.), Springer (1986), pp. 79–101.

101. G. Segal, "Elliptic cohomology," *Astérisque* **161–162**, 187–201 (1988).

102. J. P. Serre, *Algébre Locale*, Lecture Notes in Mathematics 11, Springer (1965).

103. J. Stienstra, "Formal group laws arising from algebraic varieties," *American J. Math.* **109**, 907–925 (1987).

104. S. Stolz, "Simply-connected manifolds of positive scalar curvature," *Annals of Math. (2)* **136**, 511–540 (1992).

105. S. Stolz, "Splitting certain MSpin-module spectra," *Topology* **33**, 159–180 (1994).

106. S. Stolz, "A conjecture concerning positive Ricci curvature and the Witten genus," *Math. Ann.* **304**, 785–800 (1996).

107. I. A. I. Suleiman and R. A. Wilson, "The 2-modular characters of Conway's group Co_2," *Math. Proc. Camb. Phil. Soc.* **116**, 275–283 (1994).

108. R. G. Swan, "Induced representations and projective modules," *Annals of Math.* **71**, 552–578 (1960).

109. M. Tanabe, "On Morava K-theories of Chevalley groups," *Amer. J. Math.* **117**, 263–278 (1995).

110. M. Tezuka and N. Yagita, *Cohomology of Finite Groups and Brown–Peterson Cohomology I*, Lecture Notes in Mathematics 1286, Springer, pp. 396–408 (1987).

111. M. Tezuka and N. Yagita, *Cohomology of Finite Groups and Brown–Peterson Cohomology II*, Lecture Notes in Mathematics 1418, Springer, pp. 57–69 (1990).

112. M. Tezuka and N. Yagita, "On odd prime components of cohomologies of sporadic simple groups and the rings of universal stable elements," *J. Algebra* **183**, 483–516 (1996).

113. C. B. Thomas, *Characteristic Classes and the Cohomology of Finite Groups*, Cambridge Studies in Advanced Mathematics 9, Cambridge Univ. Press, Cambridge, UK (1986).

114. C. B. Thomas, *Characteristic Classes and 2-modular Representations for Some Sporadic Groups II*, Lecture Notes in Mathematics 1474, Springer, pp. 371–381 (1991).

115. C. B. Thomas, "Elliptic cohomology of the classifying space of the Mathieu group M_{24}," *Contemporary Math.* **158**, 307–318 (1994).

116. J. A. Todd, "On representations of the Mathieu groups as collineation groups," *J. London Math. Soc.* **34**, 406–416 (1959).

117. J. A. Todd, "A representation of the Mathieu group M_{24} as a collineation group," *Annali di Matematica pura e applicata (4)* **71**, 199–238 (1966).

118. I. Turgenev, "On the Russian language," in *A New Russian Grammar* ed. by A. H. Semeonoff and J. M. Dent, 13th ed., London, p. 170 (1960).

119. E. T. Whittaker and G. N. Watson, *A Course of Modern Analysis*, 4th ed., Cambridge Univ. Press, Cambridge, UK (1927, reprinted 1969).

120. E. Witten, *The Index of the Dirac Operator in Loop Space*, Lecture Notes in Mathematics 1326, Springer, pp. 161–181 (1988).

121. E. Witten, *Mirror Manifolds and Topological Field Theory*, Essays on Mirror Manifolds, ed. S. T. Yau, International Press, pp. 120–159 (1992).

122. N. Yagita, "Cohomology for groups of $\mathrm{rank}_p G = 2$ and Brown–Peterson cohomology," *J. Math. Soc. Japan* **45**, 627–644 (1993).

123. D. Zagier, *Note on the Landweber–Stong Elliptic Genus*, Lecture Notes in Mathematics 1326, Springer, pp. 216–224 (1988).

124. O. Zariski and P. Samuel, *Commutative Algebra II*, Springer, NY (1975).

Index

GPSR Compliance
The European Union's (EU) General Product Safety Regulation (GPSR) is a set of rules that requires consumer products to be safe and our obligations to ensure this.

If you have any concerns about our products, you can contact us on

ProductSafety@springernature.com

In case Publisher is established outside the EU, the EU authorized representative is:

Springer Nature Customer Service Center GmbH
Europaplatz 3
69115 Heidelberg, Germany

www.ingramcontent.com/pod-product-compliance
Ingram Content Group UK Ltd.
Pitfield, Milton Keynes, MK11 3LW, UK
UKHW041445070726

13610UKWH00009B/20

* 9 7 8 1 4 7 5 7 8 7 5 7 3 *